D0882986

GLASS-CERAMICS

NON-METALLIC SOLIDS

A SERIES OF MONOGRAPHS

Editors

J. P. ROBERTS

Department of Ceramics, Glasses and Polymers,
University of Sheffield, England

P. POPPER

British Ceramic Research Association
Stoke-on-Trent, England

Volume 1. P. W. McMillan: GLASS-CERAMICS. 1964
(*Second edition*, 1979)

Volume 2. H. Rawson: INORGANIC GLASS-FORMING SYSTEMS. 1967

Volume 3. B. Jaffe, W. R. Cook Jr. and H. Jaffe: PIEZOELECTRIC CERAMICS. 1971

GLASS-CERAMICS

P. W. McMILLAN

University of Warwick
Coventry, Warwickshire
England

SECOND EDITION

1979

ACADEMIC PRESS

London · New York · San Francisco

A Subsidiary of Harcourt Brace Jovanovich, Publishers

ACADEMIC PRESS INC. (LONDON) LTD
24/28 Oval Road
London NW1 7DX

United States Edition published by
ACADEMIC PRESS INC.
111 Fifth Avenue
New York, New York 10003

British Library Cataloguing in Publication Data
McMillan, Peter William
Glass-ceramics. — 2nd ed. —
(Non-metallic solids).
1. Glass
2. Ceramics
I. Title II. Series
666′.15 TP862 78–73888
ISBN 0–12–485660–8

PRINTED IN GREAT BRITAIN BY
WILLMER BROTHERS LIMITED, BIRKENHEAD

PREFACE

Many developments have occurred in the field of glass-ceramics since the first edition of this monograph was published. These have included extensive research into the basic factors underlying the glass-ceramic process; studies of the relationships between the properties of glass-ceramics and their constitution; derivation of new glass-ceramic systems and substantial commercial exploitation both in specialised engineering uses and in the consumer field.

The rapid advances in the science and technology of glass-ceramics have underlined the need for an updated text. Furthermore, the great expansion of the scientific and technical literature on these materials during the intervening years provides evidence of their continuing importance and emphasises the need for an appraisal of the current state of the art.

The aim of the second edition remains essentially the same as that of the first: namely, to describe both the theoretical and practical aspects of glass-ceramics. It is hoped that the revised text will prove of value to those engaged in materials research and development, and to design engineers seeking materials with unusual and valuable combinations of properties.

Following a similar plan to that adopted in the first edition, the history of glass-ceramics is first outlined. A discussion of fundamental aspects of glass formation and structure is followed by an extensive treatment of nucleation and crystallisation processes since it is the control of these that provides the key to the successful production of high strength glass-ceramics. The important topic of metastable phase separation in glasses is discussed because of its profound influence upon crystal nucleation and growth processes. The various types of nucleating agents are discussed with regard to practical aspects and in relation to basic research on their modes of operation.

Because of their importance as tools for basic research into the materials and as aids to the control of glass-ceramic processing, methods of investigating the crystal nucleation and growth processes are discussed and illustrated by practical examples.

New information on processes for producing high strength glass-ceramics, surface-crystallised glasses and composites is included because these developments have greatly expanded the engineering potential of the materials.

The treatment of the properties of glass-ceramics has been considerably revised to include many new data and to cover additional aspects. It is believed that the relationships between properties and glass-ceramic microstructure, encompassing not only the crystal phases but also the residual glass phase, are of basic importance. For this reason, wherever possible, the properties are discussed in microstructural terms.

The discussion of the applications of glass-ceramics ranges from consideration of fields in which the materials are firmly established such as consumer products, vacumn tube envelopes and radomes to areas where the potential of the materials still awaits full exploitation, such as uses in nuclear power engineering and in the medical field.

In preparing the revised edition, discussions with my colleagues in the University and in industry have been of considerable assistance and I express appreciation of these. Appreciation is also expressed to Corning Glass Works for permission to include data on glass-ceramics Nos. 9606 and 9608 which they sell under the trademark "Pyroceram". I wish also to record my gratitude to my wife, Miss S. Callanan and Mrs P. Lewis for typing the manuscript.

Finally, I thank my wife and family for their patience and understanding throughout the period during which the revision of the monograph has been undertaken.

February, 1979 P.W.M.

CONTENTS

Chapter 1

INTRODUCTION

A. Glass-ceramics: Definition and History

Glass-ceramics are polycrystalline solids prepared by the controlled crystallisation of glasses. Crystallisation is accomplished by subjecting suitable glasses to a carefully regulated heat-treatment schedule which results in the nucleation and growth of crystal phases within the glass. In many cases, the crystallisation process can be taken almost to completion but a small proportion of residual glass-phase is often present.

In glass-ceramics, the crystalline phases are entirely produced by crystal growth from a homogenous glass phase and this distinguishes these materials from traditional ceramics where most of the crystalline material is introduced when the ceramic composition is prepared although some recrystallisation may occur or new crystal types may arise due to solid state reactions. Glass-ceramics are distinguished from glasses by the presence of major amounts of crystals since glasses are amorphous or non-crystalline.

The development of practical glass-ceramics is comparatively recent although it has long been known that most glasses can be crystallised or devitrified if they are heated for a sufficient length of time at a suitable temperature. This knowledge led to the early attempts by Réaumur, a French chemist, to produce polycrystalline materials from glass. He showed that if glass bottles were packed into a mixture of sand and gypsum and subjected to red heat for several days they were converted into opaque porcelain-like objects. Although Réaumur was able to convert glass into a polycrystalline ceramic, he was unable to achieve the control of the crystallisation process which is necessary for the production of true glass-ceramics. The materials produced by his process had low mechanical strengths and distortion of the articles during the heat-treatment process could occur.

About 200 years after Réaumur's work, research carried out at Corning Glass Works in the United States led to the development of glass-ceramics in their present form. The first important step was the discovery of photosensitive glasses. These contain small amounts of copper, silver or gold which can be precipitated in the form of very small crystals during heat-treatment of the glasses. The precipitation process occurs much more readily if the glasses are irradiated with ultraviolet light before heat-treatment and, by selective irradiation using a suitable mask or negative, a photographic image can be produced in the glass.

In later developments, it was shown that photosensitive glasses could be opacified in the irradiated regions by the precipitation of further crystals upon the original metallic crystals. Materials made in this way would not however be considered to be glass-ceramics since the crystalline material present constitutes only a minor proportion of the final material.

S. D. Stookey of Corning Glass Works made an important basic discovery when he heated a photosensitively opacified glass to a higher temperature than that normally employed in the heat-treatment process. He found that instead of melting, the glass was converted to an opaque polycrystalline ceramic material. This material had a much higher mechanical strength than the original glass and other properties such as the electrical insulation characteristics were markedly improved. The conversion from the glass to the ceramic form was accomplished without distortion of the articles and with only minor changes of dimensions. This material represented the first true glass-ceramic. Evidently, the small metallic crystals acted as nucleation sites for the crystallisation of major phases from the glass. The large number of nuclei present and their uniform distribution throughout the glass ensured that crystal growth proceeded uniformly and that a skeleton of crystals was produced to maintain the rigidity of the glass article as its temperature was raised.

The successful application of photosensitive metals as nucleation catalysts for the controlled crystallisation of glasses opened the way for the development of other types of nucleation catalysts which did not require irradiation of the glasses as a necessary step. These later methods usually depend upon the precipitation of colloidal particles within the glass to act as nucleation sites. S. D. Stookey developed a wide range of glass compositions which contained titanium dioxide as the nucleating agent. The use of metallic phosphates to promote the controlled crystallisation of glasses was discovered by McMillan and co-workers in Great Britain. Later researches by workers in a number of countries have led to the discovery of many different types of nucleating agents for glass-ceramics production.

B. The Scientific Importance of Glass-ceramics

The investigation and development of glass-ceramics are closely related to studies of nucleation and crystallisation of supercooled liquids and are therefore of general interest in this field. Glass is a very convenient medium for fundamental studies of this type because glass-like liquids have high viscosities so that the diffusion processes and atomic rearrangements which control nucleation and crystal growth occur relatively slowly. Because of the rapid increase of viscosity which occurs when the temperature falls, it is possible to arrest the crystallisation process by rapid cooling. Thus various stages in crystal growth and development can be "frozen in" permitting the use of convenient methods of examination.

Closely related to crystal nucleation and growth studies are investigations of amorphous phase separation. This subject is of interest both from the viewpoint of the basic phenomena involved and with regard to modifications of glass properties that accompany the structural change. Furthermore, the influence of prior phase separation upon glass crystallisation processes is of prime importance both with regard to glass-ceramics formation and in relation to the stability of glasses.

The wide range of compositions that can be produced in the vitreous state is particularly valuable since it allows phase transformations to be investigated in widely differing chemical environments. The development of many crystal types, including metastable and stable phases and the formation of solid solutions, can be investigated under controlled conditions. Because molten glass is a good solvent for most oxides, for certain metals and for some halides and other compounds, the effects of these, present as minor constituents, upon crystal nucleation and growth processes can be investigated. Such studies, in addition to their basic importance, are of considerable interest in relation to the development of glass-ceramic microstructures.

In addition to their value for the study of physico-chemical effects, glass-ceramics are also valuable for fundamental investigations of certain physical properties. One important field concerns the investigation of mechanical strength and fracture processes for brittle solids. Glass-ceramics are especially valuable in such studies because they can be produced to have a very fine microstructure and in addition can contain a wide variety of crystal types. A further valuable possibility is that for identical chemical compositions, the degree of crystallinity can be varied from the amorphous glass at one extreme to the almost completely crystalline glass-ceramic at the other. This latter possibility is of interest not only in studies of mechanical failure but also in the investigation of properties which are dependent on diffusion processes, such as ionic conductivity.

Basic studies on glass-ceramic systems are of interest in connection with other areas of Materials Science. In the general field they are of importance because they offer combinations of physical properties not available with other classes of materials. To the glass technologist, the development of glass-ceramics is of great interest not only because they extend the possible applications of glass-making techniques but also because the search for new glass-ceramics stimulates research into glass compositions and the relative stabilities of various types of glass. Many of these data can be of value in the development of conventional glasses and manufacturing processes. In the field of conventional ceramics it is of interest to study the relationship crystallographic constitution and physical properties. Investigations of glass-ceramics may be particularly valuable because the crystal phases present can be varied in a controlled manner and materials having identical chemical compositions but different crystallographic compositions can be prepared. The possibility of investigating the effects of variations in the proportion and chemical composition of the vitreous phase in glass-ceramics is also of value since in some conventional ceramics the vitreous phase plays an important part in determining certain properties.

Finally, the investigation of glass-ceramics is of interest to the mineralogist since materials containing unusual combinations of known crystals are possible and in addition there is the possibility of developing entirely new crystal phases which are not formed except by the devitrification of unusual glass compositions.

C. The Technological Significance of Glass-Ceramics

The process of manufacturing a glass-ceramic involves the preparation first of a glass which is shaped in its molten or plastic state to produce articles of the required form. The glass-ware is next subjected to a controlled heat-treatment cycle which brings about nucleation and crystallisation of various phases so that the final product is a polycrystalline ceramic. This method of making a ceramic material represents a radical departure from conventional ceramic manufacturing processes and it offers a number of important advantages.

Since molten glass can be obtained in a homogeneous condition, uniformity of chemical composition can easily be achieved for glass-ceramics. The homogeneity of the parent glass together with the controlled manner in which the crystals are developed results in ceramic materials having a very fine grained uniform structure free from porosity. This is beneficial in a number of ways since it favours the development of high mechanical strength and also results in good electrical insulating characteristics.

An important feature of the glass-ceramic process is that it is applicable to a wide range of compositions and this, together with the variations which can be applied in the heat-treatment process, means that various crystal types can be developed in controlled proportions. As a result, the physical characteristics of glass-ceramics can be varied in a controlled manner and this fact has an important bearing upon the practical applications of glass-ceramics. For example, the thermal expansion coefficients of glass-ceramics can be varied over a very wide range so that at one extreme materials possessing low expansion coefficients and having very good resistance to thermal shock are possible while at the other extreme materials possessing very high thermal expansion coefficients closely matched to those of common metals can be obtained.

The use of glass-working processes such as pressing, blowing or drawing offers certain advantages over the techniques available for shaping conventional ceramics since glass lends itself to the use of high-speed automatic machinery. In general, the techniques used for shaping conventional ceramics, such as extrusion, jolleying or slip-casting, are slower than glass-shaping methods and a further point is that the ceramic ware usually requires extended drying and firing periods to avoid distortion and cracking. The advantages of the glass-ceramic process are particularly apparent in the production of thin-walled hollow-ware and other shapes where the section of the material is small since unfired conventional ceramic articles of this type are fragile, while the parent glass articles in the glass-ceramic process are relatively strong.

During conversion of the glass to the glass-ceramic form, a change in dimension occurs. However, this change is small and is controllable so that control of the shape and dimensions of the glass-ceramic article can be achieved without too much difficulty. With conventional ceramics, relatively large shrinkages (40 to 50 per cent by volume) occur during the drying and firing operations and these dimensional changes may be accompanied by distortion. Consequently, control of the final dimensions is more difficult for conventional ceramics than for glass-ceramics.

The glass-ceramic process has certain special characteristics which allow new processes to be applied. Since the materials originate as glasses they can be bonded to metals by relatively simple processes based on the fact that glass in its molten state will "wet" other materials. Thus it is possible to seal the parent glass to a suitable metal and to heat-treat the composite article to convert the glass into a polycrystalline glass-ceramic. This method of producing a ceramic-to-metal seal has many advantages over the processes available for conventional ceramics which involve complicated and expensive pre-treatment and furnacing procedures.

In recent years, important advances have taken place in the control of glass-ceramic microstructures resulting, for example, in the development of machinable glass-ceramics and in the production of bulk and fibrous glass-ceramics having orientated microstructures.

Glass-ceramics have become established as commercially important materials in fields such as consumer products, vacuum tube envelopes, telescope mirror blanks, radomes for the aerospace industry and protective coatings for metals.

Glass-ceramics can be regarded as a most valuable addition to the materials available to the design engineer. Being inorganic and non-metallic they combine useful high temperature capabilities with a high degree of chemical stability and corrosion resistance. Their unique combination of properties is likely to make them attractive for a number of specialised engineering applications.

Chapter 2

CRYSTALLISATION AND DEVITRIFICATION

A. The Glassy State

1. *Glasses and Crystals*

Glass might be described as a transparent substance possessing the properties of hardness, rigidity and brittleness. Thus, with the possible exception of transparency, the properties usually thought of as characterising glass are those normally associated with solids. As we shall see, however, glass possesses a number of properties which are characteristic of the liquid state and the classification of glass as a liquid of very high viscosity rather than as a solid would be in accordance with modern views.

Various definitions of glass have been put forward but one which is widely accepted is that proposed by the A.S.T.M.: *glass is an inorganic product of fusion which has cooled to a rigid condition without crystallising.*

This definition would exclude certain organic substances such as glucose or glycerol which can be supercooled to a rigid condition without crystallising and which in this form would possess many of the characteristics of glasses. In the present volume we are chiefly concerned with inorganic glasses and their crystallisation although it is not the intention to deny the usefulness of studies of supercooling and crystallisation of organic glasses as a means of obtaining fundamental information.

The foregoing definition would also exclude amorphous substances prepared by methods other than melt cooling: for example, by vacuum evaporation techniques. Some of these materials have structures and properties that closely resemble those of glasses as defined above and therefore it may be thought artificial to exclude them from the general definition of the vitreous state.

A definition that would embrace this further class of materials is: *An amorphous substance is one in which long range order in the atomic arrangement does not exist over distances greater than 10 nm.*

According to this definition, polycrystalline solids made up of crystallines of sizes less than 10 nm would be amorphous as would a "crystalline" solid containing more than 10^{19} point defects per cm^3.

Although there is no *a priori* reason why a glass-ceramic could not be formed by the controlled crystallisation of an amorphous material prepared by one of the alternatives to melt cooling this is rarely, if ever, done in practice and we shall not therefore be concerned with this possibility.

It is useful at this point to consider the relationship between the solid and liquid states in order to obtain a clearer understanding of the type of structure exhibited by a glass. X-ray diffraction measurements for liquids show that a certain degree of regularity exists in the liquid state since the diffraction patterns consist of one or more diffuse haloes. This result implies that a liquid cannot be structureless in the same sense as a gas but that there must be some sort of grouping or arrangement of the molecules in the liquid related to that which occurs in the solid state. The units of structure (atomic or molecular groupings) are the same in the liquid as in the crystalline solid but in the liquid these units are not arranged in a regular manner. Thus the liquid possesses short-range order but not long-range order whereas the crystalline solid possesses both short-range and long-range order leading to complete regularity throughout the solid.

If a melt of a pure substance is cooled, it is generally observed that there is a definite freezing point at which solidification occurs due to the formation of crystals. However, it is sometimes possible to continue the cooling of a liquid below its freezing point without encountering crystal formation. In this case the liquid is stated to be supercooled. Supercooling is not a rare phenomenon and is especially likely if precautions are taken to exclude anything (eg dust) which can act as nuclei upon which crystals can grow. A supercooled liquid represents a metastable state since its free energy is higher than that of the corresponding crystal but the structure of the supercooled liquid has a lower free energy than any immediately neighbouring structure.

The relationship between the glassy state and the normal solid and liquid states can be understood on the basis of what happens during the cooling of melts. For a substance which crystallises, it is observed that there is a closely defined temperature at which solidification occurs and at this temperature a discontinuous volume change (often a contraction) occurs. In addition heat is evolved when solidification takes place. For a substance which can be cooled to the glass state on the other hand, no discontinuous volume change is found and there is no exothermic effect corresponding with the change from the liquid to the solid state. Instead, the viscosity of the melt increases progressively as the temperature falls and eventually the viscosity attains values which are so high that for all practical purposes the substance behaves as a rigid solid. Thus the glassy state is continuous with the liquid state and is distinguished from the normal liquid state by the high magnitude of the viscosity. On this basis, it is clear that glass may be regarded as a supercooled liquid since the melt is cooled through the temperature zone in which crystallisation might occur but the liquid-like structure is retained. With increasing degrees of supercooling, there is an increasing difference in free energy between the equilibrium solid and liquid states but the rapid rise of viscosity with fall of temperature renders the structural rearrangements necessary to permit crystallisation increasingly unlikely. Like other supercooled liquids a glass is a substance in a metastable state and it would achieve a lower free energy by crystallising. Glasses appear to be completley stable, however, except within a fairly restricted range of temperatures.

There are many properties of glass which confirm its liquid-like nature. For example, the transparency of glass may be thought of as a property more usually characteristic of the liquid state than of the solid crystalline state. It is true that non-metallic single crystals are often transparent but polycrystalline materials are not, and the transparency of glass is a result of the complete absence of grain boundaries or inclusions which could cause scattering of light. The X-ray diffraction pattern of glass shows only diffuse haloes as compared with the sharp pattern of lines given by a crystalline substance. The pattern given by a glass is thus virtually indistinguishable from that given by a liquid and this provides confirmation of the liquid-like nature of glass. Glass is isotropic; that is, its properties are the same regardless of the direction in which they are measured. Isotropy is more characteristic of a liquid structure than of a crystalline structure. The structure of glass is permeable to small particles such as sodium ions so that it behaves as an electrolytic conductor and this is associated with the rather "open" liquid-like structure. Similarly, certain glasses are permeable to the smaller gaseous atoms such as those of hydrogen and helium.

2. *Conditions for Glass Formation*

a. Structural considerations. Some oxides can readily be obtained in the form of glasses when they are cooled from the molten state whereas others invariably crystallise. Whether or not an oxide can be obtained in the form of a glass is of considerable scientific interest and it is of importance to understand the basic principles governing glass formation.

Goldschmidt (1926) made one of the earliest attempts to discover characteristics common to glass-forming oxides and suggested that the ability of an oxide to form a glass might be related to the way in which the oxygen ions were arranged around the cation to form the unit cell of the crystal structure. In stable crystal structures the number of anions immediately surrounding a cation (co-ordination number) is determined by the relative sizes of the anion and cation. It can be shown from geometrical considerations that for an oxide M_xO_y, the co-ordination number of the M cations will be four if the radius ratio R_M/R_O lies between 0·225 and 0·414. In this case, the oxygens are arranged at the corners of a tetrahedron with the cation occupying a central position. Goldschmidt pointed out that for a number of glass-forming oxides, including SiO_2, GeO_2 and P_2O_5, a tetrahedral arrangement occurred in the crystalline state and suggested that this might be a criterion of glass-forming ability.

Zachariasen (1932) pointed out that the ability of an oxide to form a tetrahedral configuration could not be an absolute criterion of glass-forming ability since the radius ratio for BeO, for example, will permit oxygen ions to form tetrahedral groupings around the beryllium ion and yet this oxide cannot be obtained in the glassy state. This led him to examine more closely the characteristics of glass-forming oxides and to develop the random network theory of glass structure.

Starting from the basis that the interatomic forces in glasses and crystals must be essentially similar and that the atoms in glass oscillate about definite equilibrium positions, Zachariasen deduced that the atoms must be linked in the form of a three-dimensional network in glass as in crystals. The network in glass could not be a periodic one since glasses, unlike crystals, do not give sharp X-ray diffraction spectra. Zachariasen also proposed that the energy content of a substance in the glassy state must not be greatly different from that of the corresponding crystal network. It follows from this that for a glass-forming oxide, the co-ordination number of the cation must be closely similar in the glass to that observed in the crystal. This means that the units of structure in the glass and in the crystal will be practically identical. In the crystal these structural units are built up to give a regular lattice but in the glass there is sufficient distortion of bond angles to permit the structural units to be arranged in a non-periodic fashion giving a random network. Figure 1 shows the differences between the regular crystalline lattice and the random network for an oxide having the formula M_2O_3. In both cases the structural unit is the MO_3 triangle. Thus glasses possess short-range order since the oxygens are arranged in fairly regular polyhedra but long-range order is absent. This view of glass structure is entirely consistent with the liquid-like nature of glass.

It should be pointed out, however, that the structure depicted in Fig. 1b represents a rather large departure from the crystal lattice in Fig. 1a. Smaller deviations in bond angles would probably be sufficient to generate a non-periodic network giving a structure sufficiently lacking in long-range order to account for X-ray and neutron scattering results. For example, in vitreous silica the variation in Si–O–Si bond angle is probably only about $\pm$ 10 per cent.

Having postulated the random network structure for glass, Zachariasen proposed certain conditions for glass formation. For an oxide M_xO_y to form a glass it was proposed that:

(*a*) an oxygen atom must not be linked to more than two M atoms,
(*b*) the number of oxygen atoms surrounding M must be small,

(*c*) the oxygen polyhedra must share corners only and not edges or faces.

A fourth condition proposed by Zachariasen, namely that at least three corners in each oxygen polyhedron must be shared, is of less importance and is not strictly applicable since glasses are known in which this condition would not be fulfilled.

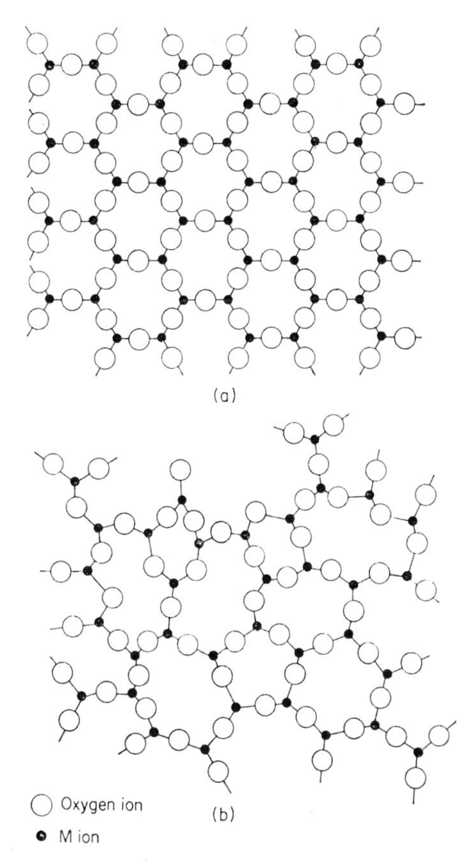

FIG. 1. Two-dimensional representation of an oxide M_2O_3 in (a) the crystalline form (b) the glassy form.

The oxides M_2O and MO cannot meet the conditions proposed by Zachariasen. The oxides M_2O_3 can do so if the oxygens form triangles around each M atom and the oxides MO_2 and M_2O_5 can do so if the oxygen form tetrahedra around each M atom.

Vitreous boric oxide (B_2O_3) is a good example of a glass whose structure is built up of triangular units; these units are found in many crystalline borates. Vitreous silica (SiO_2), germania (GeO_2), phosphorus pentoxide (P_2O_5), and arsenic oxide (As_2O_5), are examples of glasses built up of tetrahedral units; the tetrahedra SiO_4, GeO_4, PO_4 and AsO_4 are found in the crystalline state.

Oxides which form glasses when melted and cooled are called glass-forming oxides or network-forming oxides because of their ability to build up continuous three-dimensional random networks. When considering glass structure, however, it is necessary to distinguish two other types of oxide in terms of their functions in the glass structure. These are network-modifying oxides and intermediate oxides. A modifying oxide is one which is incapable of building up a continuous network and the effect of such an oxide is usually to weaken the glass network as will become clear later in this discussion. Sodium oxide is a good example of a modifying oxide. An intermediate oxide is one which although not usually capable of forming a glass, can take part in the glass network. Aluminium oxide is an example of such an oxide.

Considering first the role of a modifying oxide and taking sodium oxide as an example, we find that when this oxide is introduced into a silica glass to give sodium silicate glasses, structural changes occur as indicated in Fig. 2. Instead of the bridging oxygen ions which formed the link between two SiO_4 tetrahedra there are now two non-bridging oxygens, one of which has been contributed by the sodium oxide. Thus the effect of introducing sodium oxide has been to produce a gap in the continuous network structure. The sodium ions are accommodated in the holes or interstices in the random network structure as shown in Fig. 3. The introduction of Na_2O into the glass results in changes of the properties including reduction of the viscosity of the glass and increase in the thermal expansion coefficient. Both of these effects can be attributed to weakening of the bonds within the glass network. Other alkali metal oxides, such as lithium or potassium oxides, take part in the glass structure in a similar manner. On the average, the lithium ions will be accommodated in smaller structural interstices than the sodium ions and the potassium ions in larger ones. The alkaline earth oxides such as MgO, CaO and BaO also act a modifying oxides, the metallic cations occupying interstitial positions and the oxygen ions contributed to the glass becoming linked to the network forming ions (eg silicon). In the case of the divalent cations (Mg^{2+}, Ca^{2+}, Ba^{2+}) one cation will be present for each pair of non-bridging oxygen ions, whereas in the case of the univalent cations (Li^+, Na^+, K^+) two cations will be present for each pair of non-bridging oxygens.

Aluminium oxide is a good example of an intermediate oxide so that it is useful to consider the role of the aluminium ion in crystal and glass structures. In crystals, the aluminium ion can be four or six co-ordinated with oxygen giving rise to tetrahedral AlO_4 or octahedral AlO_6 groups. The tetrahedral groups can replace SiO_4 tetrahedra in silicate lattices to give the arrangement shown in Fig. 4. Since each aluminium ion has a charge of $+3$ as compared with a charge of $+4$ for each silicon ion, an additional unit positive charge

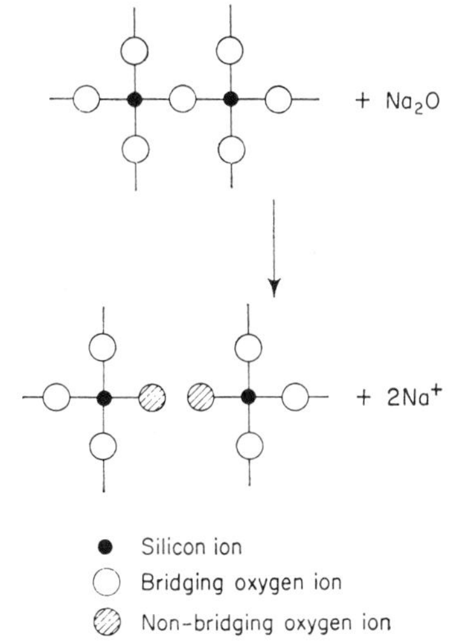

FIG. 2. Reaction between sodium oxide and silica tetrahedra. (In this figure, for simplicity, a two-dimensional representation of the SiO_4 groups is given: in the actual glass structure these groups take the form of tetrahedra.)

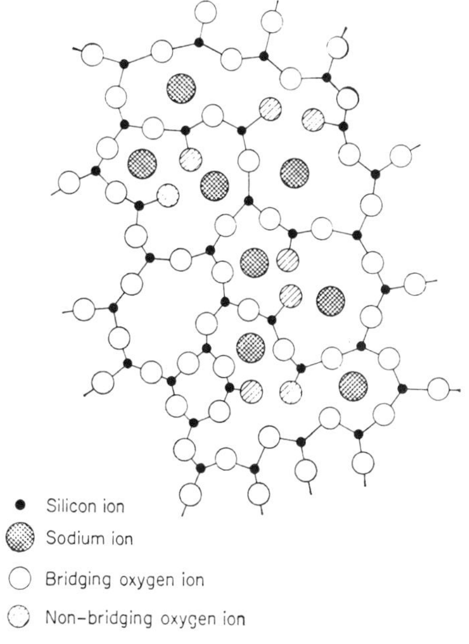

FIG. 3. Two-dimensional representation of the structure of soda-silica glass. (The structure is shown in a simplified form since only three of the four oxygen ions surrounding each silicon ion are depicted.)

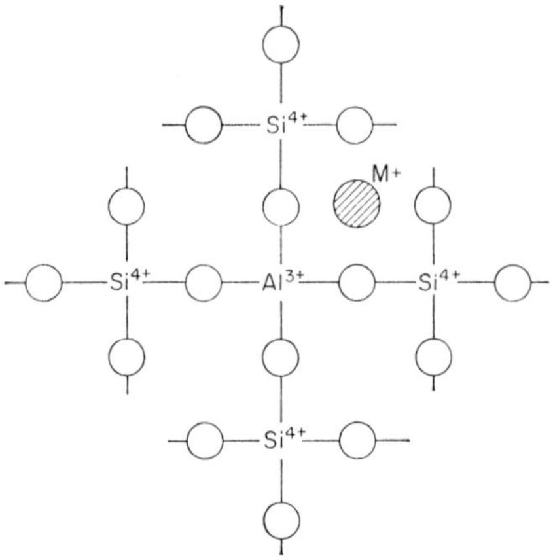

FIG. 4. Aluminium in a silicate network. (The structure is shown in a simplified form; the true structure is three-dimensional, the AlO_4 and SiO_4 groups having tetrahedral configurations.)

must be present to ensure electroneutrality. One alkali metal ion per AlO_4 tetrahedron would satisfy this requirement and the alkali metal ions could be accommodated in the interstices between tetrahedral groups. This type of structural arrangement is found for many aluminosilicates such as felspars and zeolites where the crystals are built up of linked SiO_4 and AlO_4 groups. Large univalent or divalent cations are present in these structures to the extent of one alkali ion of "half" an alkaline earth ion per AlO_4 tetrahedron. It is believed that aluminium oxide takes part in the random glass network in a similar fashion and the positive ions necessary to balance the excess negative charge on the AlO_4 tetrahedra are accommodated in the interstices of the network. The electroneutrality requirement imposes the condition that each gram-molecule of aluminium oxide present in the glass requires the presence also of one gram-molecule of an alkali oxide or alkaline earth oxide. This rule is obeyed for many types of alumino-silicate glass.

Other intermediate oxides include beryllia, BeO. In certain glasses this oxide takes part in the random network in the form of tetrahedral BeO_4 groups and two alkali metal ions per tetrahedron must be present in this case to ensure electroneutrality. Titanium dioxide and zirconium oxide may also act as intermediate oxides taking part in the glass network but the exact manner in which they do so has not been elucidated.

b. Bonding criteria. The criteria for glass formation and for characterising the roles of other oxides in glass structure discussed so far have essentially been geometrical in nature and have emphasised the role of ionic radius. It is clear, however, that the nature of the bonding between the cations and oxygen must also play a critical role. Those oxides that form highly covalent bonds to oxygen are more likely to assume the role of network formers than oxides in which the bonding is predominantly ionic.

One measure of the power of a cation to attract electrons, and therefore of ability to form covalent bonds, is the ionic field strength given by

$$F = Z/r^2$$

where Z is the valency and r the ionic radius. As will be seen from Table I, the values of F allow a classification of oxides according to their roles in glass structure, though this criterion might place Be in the category of network forming ions. The concept of field strength represents a simplified viewpoint since it considers ions to behave as rigid spheres. This is not strictly true because large ions of low charge and non-noble gas ions are deformable. Nevertheless, the field strength concept is a useful aid to understanding the roles of various ions in glass structure.

There have been other attempts to characterise oxides with regard to their ability to form glasses. One of these involved the calculation of electronegativity differences between the cation and oxygen to give a

TABLE I

IONIC FIELD STRENGTHS OF CATIONS PRESENT IN GLASSES

Ion	Ionic radius (Å)	Field strength Z/r^2	Structural role in glass
B^{3+}	0·20	75·0	Network-forming ions
P^{5+}	0·34	43·2	
Si^{4+}	0·41	23·8	
As^{5+}	0·47	22·6	
Ge^{4+}	0·53	14·2	
Be^{2+}	0·31	20·8	Intermediate ions
Al^{3+}	0·50	12·0	
Ti^{4+}	0·68	8·7	
Zr^{4+}	0·80	6.3	
Mg^{2+}	0·65	4·7	Network-modifying ions
Li^{+}	0·60	2·78	
Ca^{2+}	0·99	2·04	
Na^{+}	0·95	1·11	
Ba^{2+}	1·35	1·10	
K^{+}	1·33	0·57	

measurement of the covalent-ionic nature of the bond (Stanworth, 1950). K. H. Sun (1947) calculated the single bond energies from heats of formation and H. Rawson (1956) extended this concept to calculate the ratio between single bond energies and melting point. While all of these approaches have enabled distinctions to be drawn between oxides that can readily be obtained in the vitreous state and those that cannot, it is unfortunately true that none of them have contributed significantly to understanding of the problem of glass formation.

Advances on this front will probably require the development of a better understanding of the structure of liquid melts and of the part played by atomic arrangements and bonding in determining the kinetic factors that govern crystallisation processes. The random network theory has proved a useful concept and has provided a valuable basis for the development of ideas concerning structure-property relationships for glasses. In recent years, however, it has been suggested that the theory has limitations and that there may be a need for synthesis of ideas derived from this theory and those from the crystallite hypothesis of glass structure. The latter postulates that glass is made up of extremely small randomly orientated crystals having definite chemical compositions. This theory is evidently too much of an over simplification. Nevertheless, it seems clear that the random network theory may require modification since glass structures may not be so disordered as the theory implies. Apart from the occurrence of sub-microscopic phase separation, reviewed by James (1975), there is some evidence from high resolution electron microscopy (Freeman *et al.*, 1977) that glasses and amorphous substances may contain small domains having a higher degree of order than suggested by the random network theory. It should be borne in mind, however, that the interpretation of the electron micrographs is controversial.

c. Kinetic factors and glass formation. In the foregoing discussion, it has been implicit that the ability of a given oxide to form a glass is related to the crystal structure of that oxide. This might be questioned because the structure and properties of the melt, rather than of the solid, must surely be the governing factors. However, in many cases the short-range order in the liquid is the same as in the solid and glass formation in oxide melts probably requires a particular form of short-range order.

The difference between the melt and the crystalline solid resides mainly in the different degrees of long-range order. Crystallisation therefore requires the transformation of the liquid structure lacking long-range order into a crystalline structure where long-range order prevails. For liquids for which glass formation can occur, this transformation is relatively difficult.

The rate at which a liquid can be transformed into a crystal is given by

$$X_t = 1 - \exp\left(-\frac{\pi}{3}U^3It^4\right)$$

where X_t is the volume fraction crystallised after time t, U is the crystal growth rate and I is the nucleation rate. Thus glass formation, which requires X_t to remain below the detectable limit necessitates either U or I or both to be low. The problem of glass formation can therefore be considered from the viewpoint of factors that govern rates of crystal nucleation and growth. These factors will be considered in detail in a later section so that for the present it is sufficient to note that a high rate of nucleation will be favoured by a low value of the interfacial energy between the crystal nucleus and the liquid (or glass) phase, a high value of the volume free energy change on transforming from the liquid to the crystalline structure and a low value of the activation energy of diffusion. Similarly, a high rate of crystal growth will be favoured by a high volume free energy change resulting from the transformation to the crystalline state and a low value of the activation energy associated with atoms crossing the interface between the liquid and crystal phases (see pp. 36–38).

Uhlmann (1972) made a notable contribution towards the understanding of the influence of kinetic factors in glass formation by using time-temperature-transformation (*T-T-T*) curves to assess the probability of glass formation for a given substance. In a *T-T-T* curve, the time taken to crystallise a given volume fraction is plotted versus the undercooling ΔT. Such curves have the form given in Fig. 5 and it will be noted that there is a temperature T_N where the time τ_N, to crystallise the chosen volume fraction is a minimum. This temperature represents a critical value if crystallisation is to be avoided. Uhlmann decided to base his considerations on *T-T-T* curves derived for a

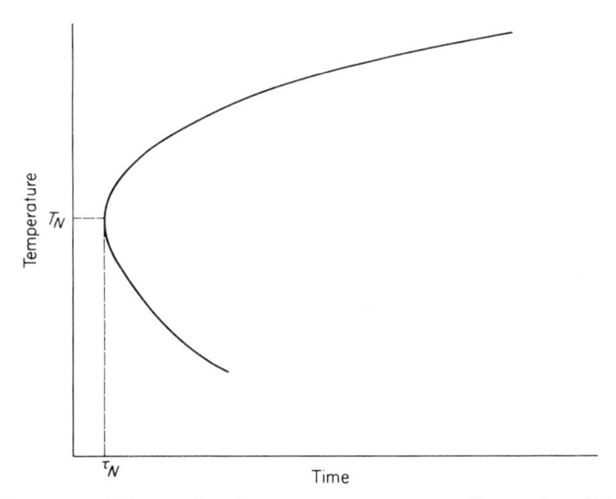

FIG. 5. The general form of a time-temperature-transformation (*T–T–T*) curve.

volume fraction $X_t = 10^{-6}$ arguing that volume fractions smaller than this could not be detected and hence materials for which $X_t < 10^{-6}$ would be vitreous.

Using either measured values of nucleation rate, I and growth rate U or values estimated by inserting suitable values in the nucleation and growth expressions, Uhlmann derived T-T-T curves for a number of materials. Two quantities that allowed an appraisal of glass-forming ability to be made were derived. These were a critical cooling rate R_c given by:

$$R_c \approx \Delta T_N / \tau_N$$

and the critical thickness y_c of material that could be cooled without the occurrence of crystallisation given by $y_c \approx (D.\tau_N)^{\frac{1}{2}}$. The values obtained for certain materials are given in Table II. It is clear that while the cooling rate for GeO_2 is well within the limits obtainable experimentally, glass formation with this oxide is appreciably more difficult than for SiO_2. The calculation indicates that formation of vitreous water requires extremely fast cooling. This is in accordance with the experimental finding that vapour deposition onto a metal plate cooled by liquid air is necessary to achieve the glassy state of water. The extremely high cooling rates that would be necessary to form amorphous or vitreous silver indicate the improbability of producing the metal in this form.

TABLE II

VALUES OF CRITICAL COOLING RATES, R_c AND CRITICAL THICKNESS, y_c. FOR GLASS FORMATION

Substance	R_c ($K sec^{-1}$)	y_c (cm)
Silica	2×10^{-4}	400
Germanium dioxide	7×10^{-2}	7
Salol	50	7×10^{-2}
Water	10^7	10^{-4}
Silver	10^{10}	10^{-5}

3. *Chemical Compositions of Glasses*

As we have seen, certain oxides, which can themselves be obtained in the glassy condition, are essential to the formation of multicomponent glasses. The most important glass-forming oxides are silica (SiO_2), boric acid (B_2O_3) and phosphorous pentoxide (P_2O_5). The great majority of commercial glasses are silicate and borosilicate compositions.

Pure silica glass is well known and is a very valuable material because its low coefficient of thermal expansion and high softening temperature render it useful where resistance to heat and thermal shock are necessary. Pure boric oxide and phosphorus pentoxide glasses, while being of scientific interest, have not found practical applications because they are both readily attacked by water.

In a multicomponent glass there will be a lower limiting proportion of the glass-forming oxide which must be present and it is of interest to consider the types of glass composition which are possible and to study some of the factors which limit the range of possible compositions. Two-component glasses can be prepared from combinations of alkali metal oxides with silica, boric acid or phosphorus pentoxide. For these compositions it is found that there is a limiting proportion of alkali metal oxide which is possible and that melts containing greater proportions of alkali oxide will crystallise or devitrify during cooling. The limiting compositions in molecular percentages for the binary alkali silicate systems are:

40 Li_2O	60 SiO_2
47 Na_2O	53 SiO_2
50 K_2O	50 SiO_2

The existence of a limiting composition is dependent on structural changes that occur as the alkali metal oxide content is increased. The introduction of non-bridging oxygen causes a progressive fall in the glass viscosity and this would favour increased crystal nucleation and growth rates. Also, if it is assumed that the non-bridging oxygens are uniformly distributed, the average number of bridging oxygens per SiO_4 tetrahedron is two, when the metasilicate composition, $R_2O.SiO_2$, is attained. At this composition therefore rings or infinitely long chains of SiO_4 groups are possible and glass-formation could occur. Further increase of the alkali metal oxide content will cause decrease of average chain lengths and disruption of rings and the conversion of the structure of the melt from a network type ultimately to one in which single SiO_4 tetrahedra are present (at the orthosilicate composition, $2R_2O.SiO_2$) will render glass formation increasingly improbable.

It is of interest to note that in the alkali metal oxide-silica systems, glass formation does not readily occur for compositions in which the average number of bridging oxygens per SiO_4 group is less than two. However, in more complex glasses the situation is somewhat different since the concentration of bridging oxygens can fall significantly below the "critical" value. For example, Moore and McMillan (1956) showed that 0·5 g melts of the molecular composition Li_2O 15, MgO 50, SiO_2 35 could be cooled as glasses. Trap and Stevens (1959, 1960a, b) have also shown that in the Na_2O–CaO–SiO_2 and Na_2O–MgO–SiO_2 systems, the SiO_2 contents can be as low as 40 molecular per cent. In such glasses, the large concentrations of non-bridging oxygens would appear to preclude the existence of large groups of SiO_4 units bonded via bridging oxygens. At a silica content of 40 mole per cent, the glass structure might consist of isolated pairs of SiO_4 tetrahedra. Trap and Stevels proposed that for glass compositions containing less than 50 mole per cent SiO_2, the role of the modifying ions changes and the ionic bonds between modifying ions and SiO_4 groups stabilise the random structure. Measurements of viscosity as a function of modifying oxide content, lent some support to this view since it was found that viscosity in the range 500° to 700°C for a complex silicate glass *increased* when the modifying oxide content was increased above 50 mole per cent, thus reversing the trend found for lower modifying oxide contents. It should be pointed out, however, that mere calculation of the average number of bridging oxygens per SiO_4 tetrahedra does not rigorously define the melt or glass structure. Clusters or chains of SiO_4 groups could be present in compositions of high modifying oxide content if the presence of isolated SiO_4 groups is also allowed, giving a structure containing anionic groups of varying sizes.

So far, the role of non-bridging oxygens introduced by modifying oxides has been considered but the nature of the added alkali metal cation is also important. Intuitively we may suppose that those modifying cations which have a high field strength and can assume low oxygen coordination numbers will have a greater tendency to order oxygen ions around themselves than larger ions. This tendency is evidenced by the greater ease with which Li_2O–SiO_2 melts devitrify as compared with other alkali–silica systems. The same effect is observed in binary borates where the limit of glass formation is 25 mole per cent R_2O in the Li_2O–B_2O_3 system and 33 mole per cent in the Na_2O–B_2O_3 system. A similar trend is observed for binary systems containing alkaline earth oxides since devitrification occurs more readily for melts containing ions of high field strength, such as magnesium, than for melts containing ions of low field strength.

In considering glasses that could be converted into glass-ceramics, we are concerned with compositions that are considerably more complex than the binary systems mentioned so far. These multicomponent systems are, however, very often modifications of a relatively restricted number of ternary systems. It is useful therefore to consider several important ternary systems that form the basis of glass-ceramic compositions.

The more complex behaviour of ternary systems is illustrated by the observation that in the Li_2O–Na_2O–SiO_2 system, glasses can be made having a greater total alkali metal oxide content than in either of the binary Li_2O–SiO_2 or Na_2O–SiO_2 systems. It is possible that the co-ordination requirements of the alkali metal ions are more easily satisfied when mixtures of ions having different radii are present.

A somewhat similar effect is found for the ternary Li_2O–MgO–SiO_2 glasses

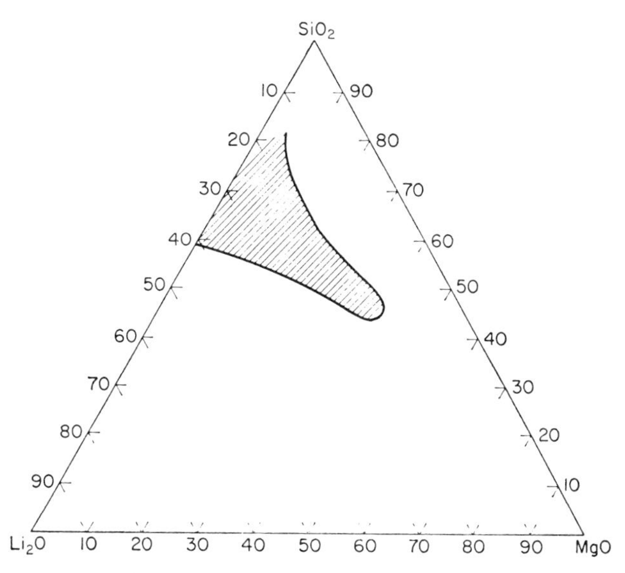

FIG. 6. Region of glass formation in the Li_2O–MgO–SiO_2 system (molecular percentage composition).

where the total $Li_2O + MgO$ content can greatly exceed the Li_2O content in the binary Li_2O–SiO_2 system. As shown in Fig. 6 the total $Li_2O + MgO$ content can be as high as 55 mol per cent. In this limiting condition the average number of non-bridging oxygen ions per SiO_4 tetrahedron would be 2·4 if all of the Li^+ and Mg^{2+} ions entered interstitial positions as network modifiers. In these circumstances, a continuous silicate network cannot exist and it seems probable that a proportion of the magnesium ions assume a network forming role. One possibility is that these ions take part in a tetrahedral MgO_4 groups in which case their behaviour would be similar to that of beryllium ions.

Aluminium oxide also reduces the tendency of binary silicate melts to devitrify and this is illustrated in Fig. 7 which shows that lithia contents up to 45 molecular per cent are possible in the ternary Li_2O–Al_2O_3–SiO_2 glasses as compared with a limiting proportion of 40 molecular per cent Li_2O in the Li_2O SiO_2 glasses. The proportion of aluminium oxide in these glasses is limited by the requirement that the ratio Al_2O_3:Li_2O cannot exceed 1:1 in order to satisfy electroneutrality conditions as discussed previously. Another important aluminosilicate system is the MgO–Al_2O_3–SiO_2 system and the limits of glass formation are given for this in Fig. 8.

The limits of glass formation for the Li_2O–ZnO–SiO_2 system (Fig. 9) are extensive and this suggests that zinc oxide may act as an intermediate oxide possibly in a similar manner to magnesium oxide.

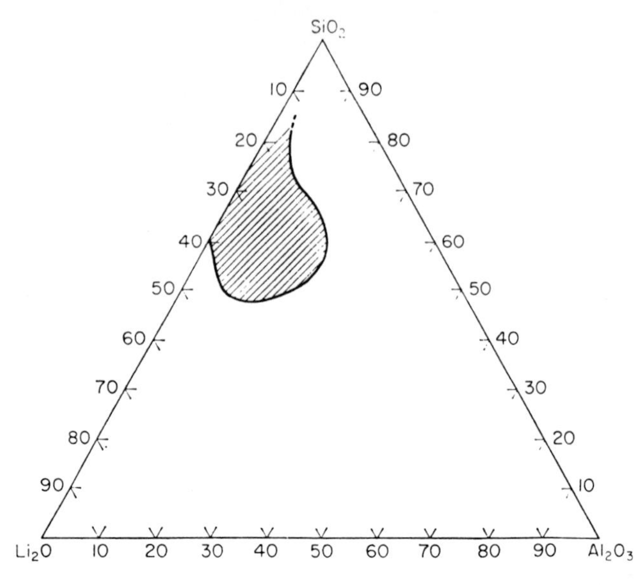

FIG. 7. Region of glass formation in the Li_2O–Al_2O_3–SiO_2 system (molecular percentage composition).

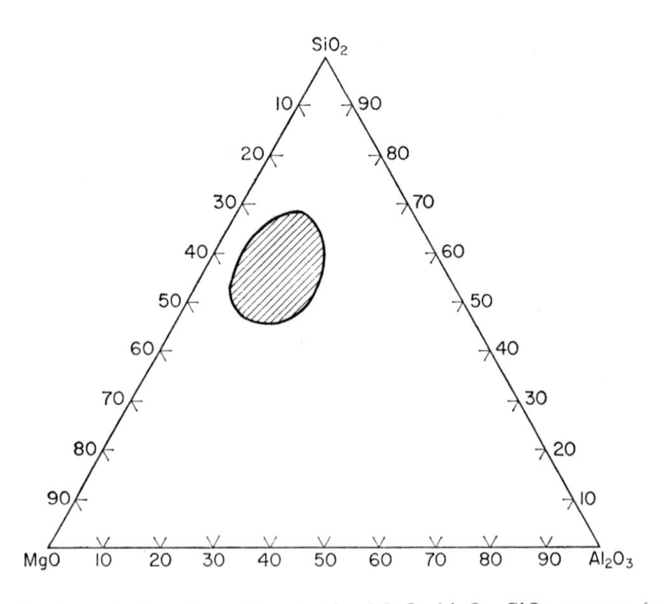

FIG. 8. Region of glass formation in the MgO–Al_2O_3–SiO_2 system (molecular percentage composition).

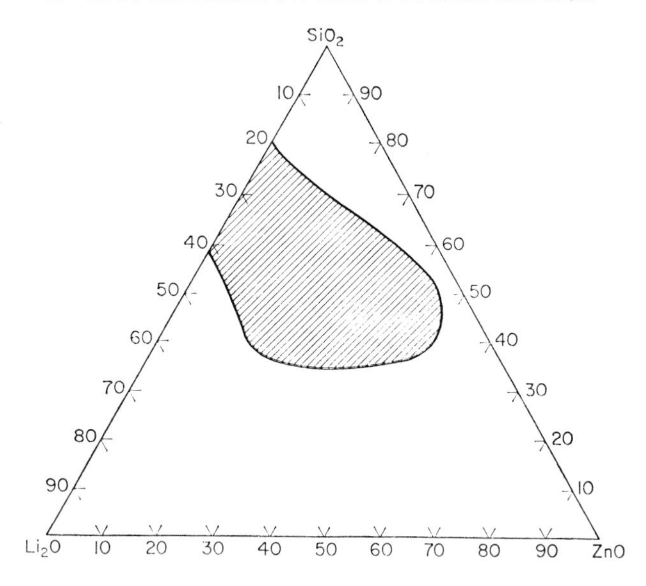

FIG. 9. Region of glass formation in the Li_2O–ZnO–SiO_2 system (molecular percentage composition).

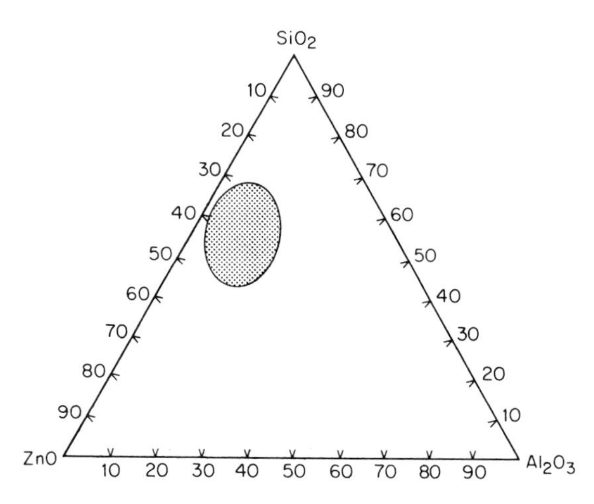

FIG. 10. Region of glass formation in the ZnO–Al_2O_3–SiO_2 system (molecular percentage composition).

The similarity between the roles of zinc and magnesium oxide is also illustrated by comparison of the glass-forming region for the $ZnO–Al_20_3–SiO_2$ system (Fig. 10) with that for the $MgO–Al_2O_3–SiO_2$ system (Fig. 8).

In addition to the oxide systems already discussed, those containing calcium oxide and lead oxide are of interest in the field of glass-ceramics. The range of binary glass compositions in the $CaO–SiO_2$ systems is restricted since for CaO contents less than about 29 molecular per cent and temperatures below 1700°C, the melt consists of two immiscible liquids. With the addition of a third component, such as alumina, the range of glass compositions is greatly extended and glasses containing high proportions (c. 45 molecular per cent) of both calcium oxide and alumina can be prepared. Lead oxide is unusual since both silicate and borate glasses containing very high proportions of this oxide can be prepared. In the silicate glasses as much as 77 molecular per cent PbO can be present and in the borate glasses the lead oxide content can be as high as 84 molecular per cent. Thus in the silicate system, glasses containing higher proportions of PbO than the orthosilicate compositions $2PbO.SiO_2$ are possible and in such glasses a continuous network of SiO_4 tetrahedra cannot be present. It is possible that the lead ions which are highly polarisable occupy "bridging" positions between adjacent SiO_4 tetrahedra and thus enable a network structure to be formed.

TABLE III

APPROXIMATE WEIGHT PERCENTAGE COMPOSITIONS OF COMMERCIAL GLASSES

Constituent	Soda-lime-silica glass	Borosilicate heat-resisting glass	Lead glass
SiO_2	70–74	80·5	56–58
B_2O_3	0–0·2	12	—
Al_2O_3	0·5–2·0	2	0–1
PbO	—	—	—
MgO	0–4	1	—
CaO	5–10		
Na_2O	12–17	4·5	
K_2O	—	—	12–13

There are many other oxide systems from which glasses can be produced, but it is beyond the scope of the present volume to deal completely with this very wide subject. The data given have therefore been mainly confined to those systems which are of the greater importance in the field of glass-ceramics. It is of interest, however, to know how the types of normal commercial glasses differ from those used for glass-ceramic production and for this reason the chemical compositions of three important types of glass are given in Table III. Such glasses have certain important characteristics including good working properties which enable them to be shaped by a variety of processes, high stability against devitrification and good chemical stability.

4. *Viscosity–Temperature Characteristics of Glasses*

Since glass is a supercooled liquid it does not have a sharp melting point but softens gradually and eventually becomes fluid due to the continuous fall of viscosity with increase of temperature. The relationship between viscosity and temperature for glasses is important in a number of respects. For example, during the melting of glasses a low viscosity favours the rapid rise of gas bubbles through the melt thus permitting clear bubble-free glass to be produced. Also the annealing of glass (to remove strains introduced as a result of uneven cooling during the shaping operations) depends upon heating the glass to a temperature where its viscosity is low enough to permit stress relief without resulting in distortion of the glass. In making glass-ceramics, which involves nucleation and crystallisation of glasses under carefully controlled conditions, the selection of optimum heat-treatment temperatures is governed by the viscosity–temperature characteristics of the glasses.

Smooth curves can be drawn relating the viscosities of glasses in poises to temperature for a wide range but it is often more convenient to define the viscosity–temperature relationship in terms of certain characteristic temperatures at which the glass attains particular values of viscosity. These characteristic temperatures are given in Table IV together with the corresponding values of viscosity.

The fibre softening point is the temperature at which a uniform fibre 0·5 to

TABLE IV
CHARACTERISTIC TEMPERATURES WITH CORRESPONDING VISCOSITIES FOR GLASSES

Characteristic temperature (°C)	Viscosity (poises)
Fibre softening point	$10^{7.6}$
Dilatometric softening point or Mg point	10^{11}–10^{12}
Annealing point	$10^{13.4}$
Strain point	$10^{14.6}$

1·0 mm diameter and 22·0 cm long elongates under its own weight at a rate of 1 mm/min when heated at a rate of 5°C per minute under standardised conditions. The dilatometric softening point (Mg point) is the temperature at which viscous flow exactly counteracts thermal expansion during measurement; it is therefore the maximum point attained on the thermal expansion curve. The annealing point is the temperature at which internal stress in glass is substantially relieved in 15 minutes. The strain point, representing the lower end of the annealing range is usually taken to be the temperature at which glass can be annealed commercially in 16 hours.

B. Crystallisation in Supercooled Liquids

1. *The Crystallisation Process—General*

Crystallisation is the process by which the regular lattice of the crystal is generated from the less well-ordered liquid structure. In its simplest form, crystallisation is observed when a melt of a single pure element or compound is cooled; conversion from the liquid to the solid state occurs at a temperature which is fixed for a given pressure and is known as the freezing point.

In the case of a salt dissolved in a suitable solvent (eg water) the situation is more complex and the conditions of equilibrium can be considered from two points of view. If the solution is in equilibrium with the solid phase of the solvent (ie ice) then the solution is said to be at its freezing point and the curve representing variation of this temperature with composition is described as the freezing-point curve. On the other hand if the solid phase of the solute (ie salt) is in equilibrium with the liquid, the latter is said to be a saturated solution and the variation of composition with temperature is represented by a solubility curve. A solution which contains greater amounts of the solute than would be expected from the equilibrium solubility curve is described as a supersaturated solution.

For mixtures of two components, A and B (eg sodium oxide and silica) which can be melted to form homogeneous liquids in all proportions ranging from pure A to pure B, the distinction between solvent and solute, and hence between freezing-point curves and solubility curves, becomes meaningless. In such cases it is usual to define the temperature composition relationship in terms of the liquidus curve. This is equivalent to a freezing point curve and can be obtained by plotting the temperatures at which crystals first appear during slow cooling of the melt against composition of the melt. Any point on this curve represents the liquidus temperature of the corresponding composition and at this temperature crystals of the primary phase (ie the first phase to appear during cooling) can exist in equilibrium with the melt. A liquid or melt which has been cooled below the liquidus temperature without the separation of crystals is described as a supercooled liquid.

Similarities exist between the crystallisation processes for supersaturated solutions and for supercooled liquid melts and we may expect that these processes will bear some relation to those which occur during the devitrification of a glass. For melts of the type which can be cooled to form glasses, the conditions relating to crystallisation are complex since there may be a large number of oxides present in mutual solution. In addition, although crystallisation may involve the separation out of only one oxide in its crystalline form, such as silica in the form of cristobalite or quartz, the crystals may be more complex, comprising at least two oxides in combination such as the various crystalline silicates. Despite the great complexity of the process, however, an understanding of the factors involved in the crystallisation of glasses can be approached by considerations of the crystallisation of simpler liquids and this is the approach which will be employed.

A very important and fundamental observation which is made for crystallisation processes, and indeed for other phase transformation processes such as the formation of liquid drops in a vapour phase, is that the transformation does not occur simultaneously throughout the mother phase. The transformation (crystallisation in the case under consideration) proceeds from distinct centres and crystal growth takes place by deposition of material upon the first tiny crystals or nuclei. Two parts of the crystallisation process are therefore distinguished: nucleation and crystal growth. Nucleation involves the formation of regions of longer range atomic order than are normally present in the liquid phase. These unstable intermediate states are known as embryos, and the embryos having a critical minimum size which are capable of developing spontaneously into gross particles of the stable phase are known as nuclei.

Nucleation may be homogeneous or heterogeneous and it is important to distinguish between the two types. In homogeneous nucleation, the first tiny seeds are of the same constitution as the crystals which grow upon them, but in heterogeneous nucleation the nuclei can be quite different chemically from the crystals which are deposited. Homogeneous nucleation in which the embryos arise due to local fluctuations in the structure of the liquid phase is often difficult to observe experimentally owing to the problem of excluding foreign nuclei such as dust particles. Generally speaking, this type of nucleation occurs at high degrees of supersaturation or supercooling of the liquid phase. It has been known for over a hundred years that certain solid bodies extraneous to the system promote crystallisation. This fact has been explained on the basis that the heterogeneities catalyse the formation of nuclei of the new phase. Different types of heterogeneities or nucleation catalysts vary in regard to their ability to promote crystallisation and the potency of the nucleation catalyst is determined by the extent to which the atomic structure of the catalyst resembles that of the phase being nucleated. In the next section, the basic aspects of nucleation will be discussed in greater detail.

2. *Nucleation and Crystallisation of Supercooled Liquids*

a. Homogeneous nucleation. For this type of nucleation all foreign nuclei must have been excluded and this is always difficult to prove. De Coppet in 1872 conducted experiments which showed that on the average the life of a supersaturated solution was inversely proportional to the degree of supersaturation and from this it may be inferred that the probability of homogeneous nucleation increases with the degree of supersaturation. Ostwald (1897) in a series of careful experiments confirmed this general effect and showed that some supersaturated solutions would last indefinitely, if foreign nuclei were excluded, without ever spontaneously forming a solid phase. These solutions are described as metastable solutions. For other supersaturated solutions, the solid phase appears spontaneously after a limited time. Such solutions are called unstable solutions. Increase of the concentration will convert a metastable solution to an unstable one and the concentration at the transition point may be called the metastable limit. For concentrations below the metastable limit, spontaneous formation of crystal nuclei evidently does not occur. For solutions in the neighbourhood of the metastable limit, spontaneous precipitation of crystals can occur when the solutions are exposed to quite small disturbing effects such as variations in pressure or temperature or local evaporation.

The existence of a metastable zone of supersaturation occurs for other phase transformation processes such as the condensation of vapours and the effect is utilised in the well-known Wilson cloud chamber where foreign nuclei such as ions or particles from radioactive disintegrations lower the limit of supersaturation of water vapour to provide a record of their tracks.

Tamman (1925) made classical studies of crystallisation in supercooled liquids including inorganic glasses which contributed greatly to the knowledge of nucleation and crystallisation processes. He showed that below the equilibrium melting temperature there exists a temperature interval, referred to as the metastable zone, in which nuclei do not form at a detectable rate. In this zone, however, crystals can grow if nuclei are provided, ie if the melt is "seeded" or "inoculated". At temperatures below this region the crystallisation process is controlled by two factors: the rate of formation of nuclei and the crystal growth rate. Melts which increase rapidly in viscosity during cooling, such as those which can form glasses, show maxima in nucleation and crystal growth rates because at the lower temperatures the high viscosity hinders the atomic rearrangements and diffusion processes which are

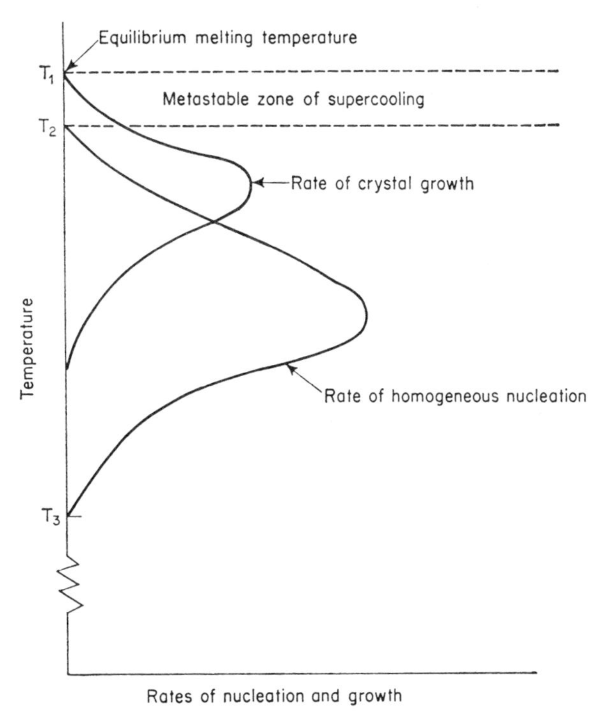

FIG. 11. Rates of homogeneous nucleation and crystal growth in a viscous liquid.

necessary for nucleation and crystal growth. Consequently, curves for nucleation rate and crystal growth rate for a viscous melt have the form shown in Fig. 11.

It is evident from this figure that if the aim is to produce the largest possible number of small crystals, nucleation should occur at or near to the temperature at which the maximum nucleation rate occurs. As will become clear from later chapters, selection of the optimum nucleation temperature is important in the production of glass-ceramics.

The metastable zone of supercooling ($T_1 - T_2$) below the equilibrium melting temperature occurs because the very tiny crystal nuclei have melting temperatures appreciably below that of the bulk material. Similarly, the metastable zone of supersaturation in solutions results from the higher solubility of small crystallites as compared with that of larger crystals. From Fig. 11 it will be noted that there also exists a temperature, T_3, below which the homogeneous nucleation rate is zero due to the high viscosity of the melt.

Theoretical studies of nucleation of supercooled liquids usually deal only with condensed systems where the vapour pressures of the solid and liquid phases are negligibly small compared with atmospheric pressure. In order to form a new phase (the nuclei) within the mother phase, an increase in free energy of the system must occur and this constitutes a barrier to nucleation. J. Willard Gibbs (1928) studied this problem and derived an expression for W, the work done to form a spherical mass of a new phase B within the mother phase A. The expression has the following form:

$$W = 16\pi\sigma^3/3(p''-p')^2 \tag{1}$$

where σ is the interfacial tension, p' is the pressure within phase A and p'' is the pressure within phase B.

There are two contributions to the change of free energy that occurs on nucleation. First, the formation of a boundary or surface between the embryo and the mother phase results in a gain of free energy due to the interfacial energy. Secondly, since the arrangement of the atoms within the embryo will be less disordered than that in the surrounding phase, there will be a reduction in free energy of the system. There are thus two opposing factors which govern the actual free energy change and these can be approximated in the following fashion to give ΔF, the free energy change for a spherical inclusion of radius r:

$$\Delta F = -\tfrac{4}{3}\pi r^3 \Delta f_v + 4\pi r^2 \Delta f_s \tag{2}$$

where Δf_v is the change of free energy per unit volume resulting from transformation from one phase to the other and Δf_s is the energy per unit area of surface between the two phases; Δf_s can be equated with σ, the interfacial tension. Under certain circumstances it would be necessary to include a term in the right-hand side of the expression for the energy increase due to the strain energy resulting from the change in atomic arrangement within the embryo as compared with the surrounding phase. In the case of a supercooled melt, however, this effect would normally be negligible. Examination of Eqn (2) shows that when r is small the interfacial energy term predominates, but with increase of radius of the embryo the interfacial energy becomes a smaller fraction of the total energy change. Once some critical radius is reached, the volume free energy term will predominate and further growth will lead to a lower free energy and therefore to a more stable system. Regions smaller than the critical radius (embryos) require an increase in free energy to form and they will continually form and redissolve with the total number in the system remaining constant. Some embryos will attain the critical radius and will continue to grow with a decrease of free energy and hence will constitute stable nuclei. At the critical radius, the free energy change will attain its maximum value and by differentiation of Eqn (2) the critical radius r^* may be derived:

$$r^* = 2\,\Delta f_s/\Delta f_v \tag{3}$$

This corresponds with a maximum free energy change ΔF^* as given by

$$\Delta F^* = 16\pi(\Delta f_s)^3/3(\Delta f_v)^2 \tag{4}$$

which is, of course, similar to the Gibbs Equation (1).

Becker and Doering (1935), Frenkel (1946) and others have employed statistical mechanics to derive expressions for I, the rate of homogeneous nucleation. These have the form:

$$I = A \exp(-\Delta F^*/kT) \tag{5}$$

where A is a constant and ΔF^* is maximum free energy of activation for formation of a stable nucleus as given above.

This expresssion neglects the effects of diffusion rate but in a viscous liquid the activation energy for diffusion of molecules across the phase boundary may well constitute a major barrier to nucleation. Therefore, a general equation for the rate of homogeneous nucleation in condensed systems proposed by Becker (1938) gives a better approach. This expression is:

$$I = A \exp\left[\frac{-(\Delta F^* + Q)}{kT}\right] \tag{6}$$

where Q is the activation energy for diffusion of molecules across the phase boundary. This theory is approximate only but it does at least serve to provide a qualitative explanation of experimental observations. For small degrees of supercooling, ΔF^* is large since the value of the volume free energy change Δf_v is very small and consequently the nucleation rate is low. With further supercooling Δf_v increases markedly until ΔF^* becomes comparable in magnitude with Q; under these conditions the maximum nucleation rate is achieved. The nucleation rate diminishes with further supercooling when ΔF^* becomes negligible in comparison to Q.

It is possible to develop Eqn (5) by including parameters that can in general be measured and hence to offer the prospect of estimating nucleation rates. The volume free energy change, Δf_v, in Eqn (3) is given by

$$\Delta f_v = \mathrm{L}\Delta T/VT_m \tag{7}$$

where L is the latent heat of fusion, ΔT is the undercooling below the liquidus temperature T_m and V is the gram atomic volume of the crystal. The pre-exponential term in Eqn (5) is equal to $n\upsilon$ where n is the number of atoms per cm^3 and υ is the vibrational frequency at the nucleus-liquid interface. Also, the diffusion coefficient, D, for transport of molecules across the phase boundary is given by:

$$D' = a_0^2\ \upsilon \exp(-Q/kT) \tag{8}$$

where a_0 is the interatomic distance.

Hence,

$$I = \frac{nD'}{a_0^{\,2}} \exp\left(\frac{-16\pi(\Delta f_s)^3 V^2 T_m^{\,2}}{3L^2 \Delta T^2 kT}\right) \tag{9}$$

If the Stokes-Einstein relation is assumed to hold so that

$$D' = RT/N3\pi\eta a_0 \tag{10}$$

the diffusion coefficient D' is inversely related to η, the viscosity. Thus the effect of the pre-exponential term in Eqn (8) would be to cause a rapid fall in the nucleation rate when the melt temperature is reduced below T_m. At the same time, however, the value of the exponential term increases because of the increase of the undercooling ΔT. The net result is to cause a maximum in the nucleation rate as observed experimentally.

The foregoing treatment has been concerned with steady state nucleation rates, that is with the linear part of the curve relating number of nuclei at a fixed temperature with time shown schematically in Fig. 12.

It will be seen, however, that there is an initial period during which the nucleation rate is either zero or very small and that for higher nucleation times, a saturation effect occurs wherein the number of nuclei per unit volume becomes almost constant.

In the region where saturation effects have not become important, the nucleation rate after time t is given by:

$$I(t) = I_0 \exp\,(\tau/t) \tag{11}$$

where τ is the incubation time and I_0 is the steady state nucleation rate.

b. Heterogeneous nucleation. Fahrenheit, in the eighteenth century, discovered that water could be successfully supercooled in some closed vessels

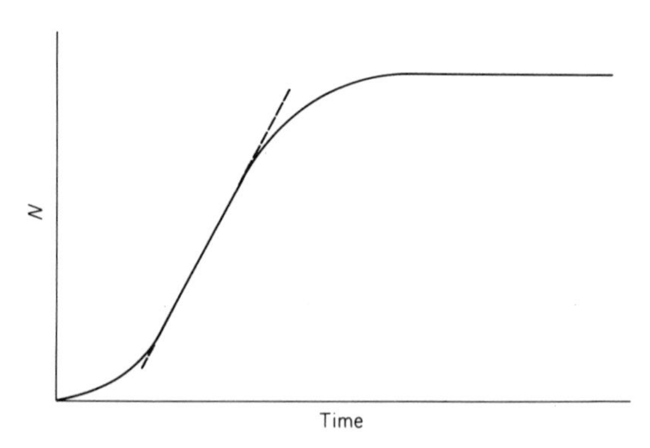

FIG. 12. Number of nuclei per unit volume, N, as a function of time at a fixed temperature.

but not in others. Inoculation with ice crystals or sometimes simply opening the vessels would cause crystallisation. Gernez, in the nineteenth century, showed that while crystallisation could be induced in supersaturated solutions by seeding with "foreign" crystals, these were less effective than crystals of the actual phase to be precipitated although isomorphous crystals were more "potent" than non-isomorphous crystals.

More recently, Vonnegut (1947) seeking means for "seeding" clouds to induce rainfall selected silver iodide as likely to be a potent nucleating agent since the lattice parameters of this crystal closely resemble those of ice. Water vapour seeded with silver iodide precipitated ice crystals with supercooling of only 4 to 6°C whereas other substances, such as atmospheric dusts, whose structures deviate markedly from that of ice, required supercooling of 20 to 30°C in order to initiate the phase change.

Vonnegut (1948) also studied the crystallisation of supercooled metal melts. He realised that sub-division of the metal melt into small droplets would overcome two serious experimental difficulties in such studies. First, it is practically impossible to prepare liquids which are free from foreign particles that might act as nuclei. Secondly, once a supercooled liquid has been nucleated, crystal growth occurs very rapidly. When the melt is sub-divided into very small droplets, only a few of these will contain a nucleating particle and thus most of the drops will have to solidify by homogeneous nucleation. Vonnegut showed that for tin spheres separated with an oil film, supercooling of 110°C could be achieved whereas for the bulk metal it is not possible to supercool by more than 20 to 30°C.

Holloman and Turnbull (1951, 1953) and Turnbull and Cech (1950) established values for the homogeneous nucleation temperatures of a number of metals using a technique in which they studied isolated drops on a quartz plate. The extent of supercooling before the onset of homogeneous nucleation varied quite widely from 77°C for mercury to 370°C for platinum. In fact for metals, a simple relationship between the supercooling, ΔT_c, and the melting temperature, T_m, seems to exist, as shown by Buckle (1960) such that $\Delta T_c/T_m \approx 0{\cdot}18$. Other substances as diverse as alkali halides and molecular compounds show similar behaviour.

Turnbull (1952) was able to demonstrate the effects of various surface coatings on the supercooling of mercury droplets. In the case of droplets coated with mercury laurate, supercooling of 77°C was observed indicating no catalysing effect of the coating (ie homogeneous nucleation). A mercury iodide coating caused reduction of supercooling to 48°C but mercury sulphide was even more effective in promoting nucleation (supercooling 12°C). The greater potency of the latter as a nucleation catalyst was attributed to the close similarity in lattice spacings between mercury sulphide and solid mercury. These observations indicate that the potency of a nucleating agent is dependent on the relationship between the structures of the catalysing particle and of the phase being nucleated. Further evidence on this point comes from observations made during the casting of metals such as aluminium where it is found that small additions of other elements (eg titanium) lead to a marked refinement of the grain size of the cast metal. There are indications that effective grain refining occurs when the disregistry between similar low index planes of the nucleation catalyst and the metal is less than 10 per cent.

This question of the permissible disregistry between the lattice spacings of the nucleation catalyst and the phase being nucleated is an important one and consideration of the phenomenon of oriented overgrowth or epitaxy is relevant to the problem. In epitaxial growth, crystals form on the surface of a foreign crystal with a definite orientation and it is generally believed that the effect arises from oriented nucleation on the catalyst surface. In many cases, the criterion which determines whether epitaxial growth occurs is that disregistry between similar low index planes for the foreign crystal and the precipitated crystal should not exceed 10 to 20 per cent. Cases are known, however, where epitaxial growth has occurred for much larger disregistry. It seems to be generally agreed that to be effective, nucleation catalysts for the crystallisation of supercooled melts or supersaturated solutions should have a maximum disregistry of 15 per cent for spacing in low index planes of the catalyst and the nucleated phase. With increasing mismatch of crystal lattice, greater degrees of supercooling will occur before nucleation takes place.

The effect of pre-existing surfaces (such as colloidal inclusions, container walls, etc.) in a supersaturated solution or a supercooled melt is to reduce the value of ΔF^*, the free energy for homogeneous nucleation, by decreasing the value of Δf_s, the interface energy which appears in Eqn (4). The volume free energy change Δf_v between the liquid phase and the crystal phase is not altered by heterogeneous nucleation and neither is the activation energy for diffusion, Q. The important feature of heterogeneous nucleation is that the interfacial tension between the heterogeneity and the nucleated phase must be low. Therefore the influence of the catalysing surface is determined by the contact angle θ at the substrate-melt-precipitate junction. The activation energy of heterogeneous nucleation can be expressed as:

$$\Delta F_c^* = \Delta F^* f(\theta). \tag{12}$$

Turnbull and Vonnegut (1952) modified the equation for homogeneous nucleation rate to give an expression for heterogeneous nucleation rate:

$$I_c = A^1 exp\ [-\Delta F^* f(\theta)/kT] \tag{13}$$

and if the activation energy for diffusion is included, as suggested by Stokey (1959a), the equation becomes:

$$I_c = A^1 \exp\ [-(\Delta F^* f(\theta) + Q)/kT]. \tag{14}$$

Thus heterogeneous nucleation can be described in terms of the parameters which describe homogeneous nucleation and a single additional parameter, θ. The term $f(\theta)$ is given by:

$$f(\theta) = (2+\cos\,\theta)(1-\cos\,\theta)^2/4 \qquad (15)$$

if the nucleus has the form of a spherical cap on the heterogeneity; the latter is assumed large compared with the nucleus. The angle θ is determined by the balance of surface tensions:

$$\sigma_{HL} = \sigma_{SH} + \sigma_{SL}\cos\,\theta \qquad (16)$$

where:

σ_{HL} is the interfacial free energy (tension) between the heterogeneity and the melt,

σ_{SH} is the interfacial free energy between the heterogeneity and primary crystal phase, and

σ_{SL} is the interfacial free energy between the crystal phase and the melt.

For any contact angle (θ) less than 180°, the free energy barrier is less for nucleus formation on the surface of the heterogeneity than for homogeneous nucleation. As a result, heterogeneous nucleation will occur wherever possible in preference to homogeneous nucleation.

The presence of heterogeneities which have the effect of reducing ΔF^*, will also reduce the incubation time τ. For heterogeneous nucleation the incubation time τ_H is given by

$$\tau_H = \tau f(\theta) \qquad (17)$$

where τ is the incubation time for the "pure" glass, and $f(\theta)$ is given by:

$$f(\theta) = (1-\cos\,\theta/2)-(\cos^2\theta/2) \qquad (18)$$

With increasing efficiency of the nucleating particle, the incubation time is reduced and the potency of a nucleation catalyst can be expressed as τ/τ_H. Gutzow and Toschev (1971) used this ratio to compare the efficiencies of a number of metallic nucleation catalysts in promoting the crystallisation of $NaPO_3$ from a Na_2O–P_2O_5 glass. From their study it appeared that the metal catalyst investigated could be placed in order of increasing potency as follows:

gold → silver → rhenium → palladium → platinum → iridium.

There was, however, no apparent correlation between the lattice disregistry and the efficiency of catalysation. However, the disregistries did not exceed 5 per cent for any of the metals and thus this factor was well within the limit accepted to allow epitaxial growth. Thus, as observed, the predominating factor was the value of the angle of contact θ between the metal substrate and the nucleated crystal phase. The values of $f(\theta)$ in fact decreased from 0·125 for gold to 0·007 for iridium in line with the increased potency of nucleation catalysis.

The same authors (*loc. cit.*) reported that the efficiency of nucleation catalysis in $Na_2B_4O_7$ glass increased in the order Au → Ir → Pt and in this case also lattice disregistry appeared to be of minor importance.

To summarise the considerations of heterogeneous nucleation: there are two important criteria which determine the effectiveness of a nucleation catalyst. These are:

(*i*) the interfacial tension between the nucleation catalyst and the primary crystal phase must be low;

(*ii*) the crystal structures must be closely similar so that the disregistry on low index planes is not more than about 15 per cent.

c. Crystal growth. The controlled crystallisation of glasses involves crystal growth as well as nucleation. While the latter process is highly critical to the production of microcrystalline glass-ceramics, the growth process is also of considerable importance in determining the morphology of the material produced.

Crystal growth is dependent upon two factors:

(*i*) the rate at which the irregular glass structure can be re-arranged into the periodic lattice of the growing crystal;

(*ii*) the rate at which energy released in the phase transformation process can be eliminated that is: the rate of heat flow away from the crystal-glass interface.

Considering first the process of structural transformation: an atom in the crystal structure has a lower free energy than a corresponding atom in the glass phase by an amount ΔG, the bulk free energy of crystallisation. Also, for an atom to cross the interface between the glass and crystal phases it must overcome an energy barrier corresponding to a free energy of activation, $\Delta G''$.

Thus an expression for crystal growth rate, U, can be derived:

$$U = a_0 \upsilon \exp [-\Delta G''/RT] [1 - \exp \Delta G/RT] \qquad (19)$$

where a_0 is the interatomic separation and υ is the vibrational frequency at the crystal-glass interface.

The expression describes what is known as normal growth. It is based on the assumption that the probability of an atom being added to or removed from a given site is the same for all sites on the crystal-liquid interface and this requires the interface to be rough on atomic scale. In fact this is not generally true; growth generally takes place by mechanism in which it is energetically favourable for atoms to be added to step sites provided by screw dislocations intersecting the interface. Thus only a fraction of the total number of sites is available for growth; this fraction is given by:

$$f \approx \Delta t/2\pi T_m \tag{20a}$$

It is necessary therefore for the right-hand side of Eqn (19) to be multiplied by this factor. For small undercoolings, f is small but increases for large undercoolings.

Another model for crystal growth, known as surface nucleation growth, involves first the formation of a two-dimensional nucleus on the interface and the growth of this to form a monolayer. This model implies an atomically smooth interface; generally this type of behaviour is not found for silicate melts and glasses.

Returning to Eqn (19), a further expression can be derived as follows. The diffusion coefficient, D, is given by:

$$D = a_0^2 \, \upsilon \exp[-\Delta G''/RT] \tag{20b}$$

and also,

$$D = RT/3\pi N\eta a_0 \tag{21}$$

Substituting these and introducing the factor f we have

$$U = fRT\,[1 - \exp \Delta G/RT]/3\pi N a_0^2 \eta \tag{22}$$

This expression makes clear the marked effect of change of viscosity in modifying crystal growth rates in glasses. Thus although the growth rate increases at first as the melt is cooled below the liquidus temperature, the rapid increase of viscosity soon exerts a dominating influence causing the growth rate to fall and giving rise to the typical growth rate curve (Fig. 11).

A different approach, which also emphasises the importance of viscosity, starts by considering the growth rate as being given by the gradient in driving force for crystallisation divided by the unit force required to bring about rearrangement of the liquid to crystalline structure. The gradient in driving force is given by the bulk free energy ΔG divided by the interatomic separation a_0.

The force to move a particle of diameter d through a medium of viscosity η is:

$$F = 3\pi\eta d \tag{23}$$

Hence growth velocity is given by:

$$U = \Delta G/3\pi\eta a_0 d \tag{24}$$

and if $a_0 \approx d$ and $\Delta G = \Delta H f \Delta T/T_m$,

$$U = \Delta H f \Delta T/3\pi\eta a_0^2 T_m \tag{25}$$

Hence,

$$U = \text{Const. } \Delta T/\eta \tag{26}$$

This relationship is found to hold for a number of glass-formers such as SiO_2, P_2O_5 and GeO_2 and also for more complex glasses of the Na_2O–CaO–SiO_2 type but it is less reliable when applied to binary alkali silicate glasses for small undercoolings.

As pointed out earlier, in some cases a further factor controlling crystal growth rate may be the rate at which heat can flow away from the growing crystal (Doremus, 1973). This situation occurs when the rate of heat flow is unsufficient to prevent an increase of the temperature of the crystal-liquid interface to a value which will cause a reduction of growth rate. In general two factors are important: ΔHf the molar heat of crystallisation and the thermal diffusivity of the liquid surrounding the crystal.

The heat flux J created during crystallisation is:

$$J = UA\rho\Delta Hf/M \tag{27}$$

where U is the linear growth rate, A is the interfacial area, ρ is the density, and M is the molecular weight.

Also, the heat flux is proportional to the temperature gradient,

$$J = AKdT/dx \tag{28}$$

where K is the thermal conductivity.

Hence,

$$dT/dx = UP\Delta Hf/MK \tag{29}$$

For some silicate glasses the maximum growth rates do not exceed 1 μm sec^{-1} and inserting this and suitable values for other parameters into Eqn (29) gives a value for the temperature gradient of 3°C cm^{-1}. Thus it would appear that crystal growth rates in these glasses are not significantly affected by heat flow considerations. In certain glass-ceramic systems crystal growth rates as high as 10 μm sec^{-1} are observed, and here the temperature gradients at the crystal glass interface may be sufficient to have an effect upon crystal growth rates.

C. Nucleation and Crystallisation of Glasses

1. *Factors Determining the Stability of Glasses*

As we have seen earlier, all glasses are metastable substances and for ordinary glasses the equilibrium state at room temperature would be the crystalline state. The failure of glasses to crystallise when they are cooled from their melting temperatures raises questions of great scientific importance and consideration of the factors which govern the stability of glasses is relevant to the field of glass-ceramics.

There are two possible reasons for the failure of supercooled melts to crystallise. Either the nucleation rate is very small or the rate of growth of crystals upon the nuclei is negligible at all temperatures. It is not easy to decide which of these factors is the more important and there is no apparent reason why nucleation and growth rates could not be equally important, one parameter perhaps being the dominant factor for some glasses while the other may be more important in other cases.

It is generally supposed that heterogeneous nucleation is unlikely to occur during the cooling of normal glasses because molten glass is a good solvent for most types of accidental "dirt". We may regard this belief as reasonable because presumably the most potent nucleation catalysts for the formation of crystalline silicates would be substances which are close in crystalline structure to the silicates and would therefore in all probability be silicates themselves. These would, of course, be rapidly dissolved by the glass-melt. Heterogeneous nucleation can be the precursor to devitrification of glass if the foreign particles are brought into contact with the glass under conditions where they would not be dissolved. Dust falling onto glass which is subsequently reheated might, for example, provide heterogeneous nucleation.

Homogeneous nucleation of glass-forming liquids is also thought to be improbable because the activation energy of diffusion, Q in Eqn (6), is relatively high even at high temperatures and increases as the temperature falls. Under these circumstances, it is possible to cool the melt through the "dangerous" temperature zone in which nucleation might occur without a sufficient number of stable nuclei developing. Of course the time the glass spends in the critical zone of temperature would be important in determining whether nucleation could occur, but usually glasses would be cooled quite rapidly through this zone.

If crystal growth rate were the outstanding factor in determining the stability of a glass, assuming that homogeneous nucleation could occur, we should still expect that it would be possible to cool the glass through the critical temperature range without encountering devitrification since the crystal growth rate, like the nucleation rate, falls to zero at a temperature where the viscosity has become so high as to prevent the diffusion processes which control crystal growth. Below this temperature the glass is for all practical purposes completely stable.

Thus whether nucleation rate or crystal growth rate is regarded as being the paramount factor governing glass stability, it is the rapid increase of viscosity with falling temperature which effectively prevents crystal formation. For this reason, useful silicate glasses have high viscosities at the liquidus temperature, the temperature at which the primary crystalline phase can exist in equilibrium with the liquid glass. Low liquidus temperatures favour the stability of glass and commercial soda-lime silica glasses have liquidus temperatures in the region of 900 to 1000°C. These compositions lie in the phase field where devitrite ($Na_2O.3CaO.6SiO_2$) is the primary phase, and, of course, an essential requirement for forming a satisfactory glass is that this compound shall not crystallise out during the cooling of the glass.

2. *Glass-in-Glass Phase Separation*

a. General observations. It has been observed for many glass-forming systems that subjection to thermal treatment can cause separation into two non-crystalline phases. If this phase separation takes place in the melt at a temperature above the liquidus temperature, it is described as stable immiscibility whereas phase separation occurring below the liquidus is described as metastable immiscibility or glass-in-glass phase separation. A general review of liquid immiscibility in glass-forming systems has been given by James (1975).

Clearly, if a phase-separated microstructure develops in the glass prior to crystal nucleation and growth, it is likely that the kinetics of these processes will be significantly affected.

In glass-ceramics, as a general rule, metastable immiscibility is probably the more important process. It should be noted that there are two possible routes for phase separation: by a nucleation and growth process or by spinodal decomposition. The theory for the former parallels closely that for crystal nucleation and growth. The basic principles of spinodal decomposition and the kinetics of this form of phase separation with reference to glasses were elucidated by Cahn and co-workers (1961, 1965).

Figure 13a illustrates the free energy versus composition curve for a hypothetical mixture of two components A and B at temperature T. The presence of two minima in this curve creates a situation in which the mixture will separate into two phases for composition lying between U and V. Clearly the free energy of the mixture, given by a point lying on the common tangent to points U and V on the, free-energy curve, will be reduced if the mixture undergoes phase separation. At low temperatures the two free-energy minima are widely separated, but move closer together as the temperature increases. Thus the loci of the minima trace out an immiscibility dome as shown in Fig. 13b. Ultimately, the two minima coincide at a temperature T_C, known as the upper consulate temperature.

Within the immiscibility dome two distinct regions arise. Points x and y represent inflections in the free-energy curve such that between x and y $\delta^2F/\delta C^2$ (the second differential of the free energy versus composition relationship) is negative. In this region therefore a small fluctuation in

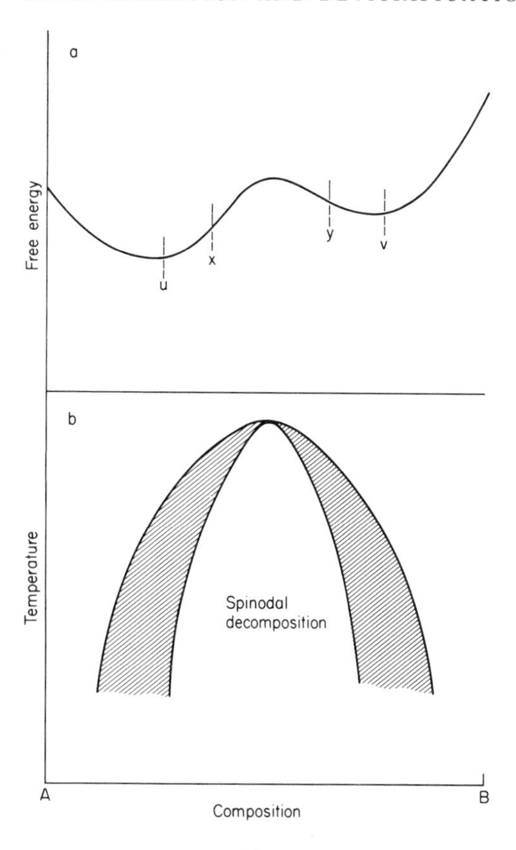

FIG. 13. Free energy versus composition relationship and the resultant phase diagram showing the regions of phase separation by spinodal decomposition and by nucleation and growth (shaded).

composition will be associated with a reduction of free energy; hence such a fluctuation will be stable and will tend to grow. Phase separation within this region, which does not require a nucleation process, is referred to as spinodal decomposition. As the temperature is raised the points of inflection (x and y) will move closer together and again will coincide at the consulate temperature. Thus, the loci of these points will trace out an inner dome as shown in Fig. 13b; the region within this is known as the spinodal because of its supposed spine like form. Referring again to Fig. 13a, in the regions between U and x and between V and y $\delta^2 F/\delta C^2$ is positive and here a small composition fluctuation will be associated with an increase of free energy. Such fluctuations will therefore be unstable and tend to redissolve. They will only continue to grow if they attain a critical size. Hence phase separation within the region outside the spinodal curve will be by a nucleation and growth process.

The essential difference between phase separation produced by the two processes can be summarised as follows: in the nucleation and growth process the interface between the two phases is always sharply defined whereas in spinodal decomposition the interface is diffuse initially and only becomes distinct at a later stage of the process; also with the spinodal decomposition the composition of the second phase changes throughout the process while it is constant for the nucleated phase; finally, in spinodal decomposition the phase-separated regions are spaced apart in a regular manner while nucleation and growth leads to irregular spacing. Table V summarizes the differences between the two phase-separation processes.

TABLE V

CHARACTERISTIC FEATURES OF PHASE SEPARATION BY NUCLEATION AND GROWTH OR BY SPINODAL DECOMPOSITION

	Nucleation and growth	Spinodal decompostion
A. Phase compositions	At a fixed temperature the nucleated phase composition remains constant	Compositions change until equilibrium is achieved
B. Interface between phases	Clearly defined throughout	Initially diffuse but ultimately becomes sharp
C. Morphology	The nucleated phase generally appears as spherical particles random in size and spacing; low connectivity	The second phase is generally non-spherical, "threadlike", with regular spacings and dimensions; high connectivity

The precise differences between the effects upon crystal nucleation that might result from prior phase separation of the two forms are not well understood but it seems possible that phase separation by nucleation and growth may play a more important role. In this the phase-separated particles have a clearly defined composition and would be more likely to crystallise by homogeneous nucleation and this could allow heterogeneous nucleation of crystallisation in the surrounding glass phase.

Examination of the phase diagram in Fig. 13 will indicate that, in general, cooling of a glass from a temperature above the immiscibility boundary will cause the region of phase separation by nucleation and growth to be

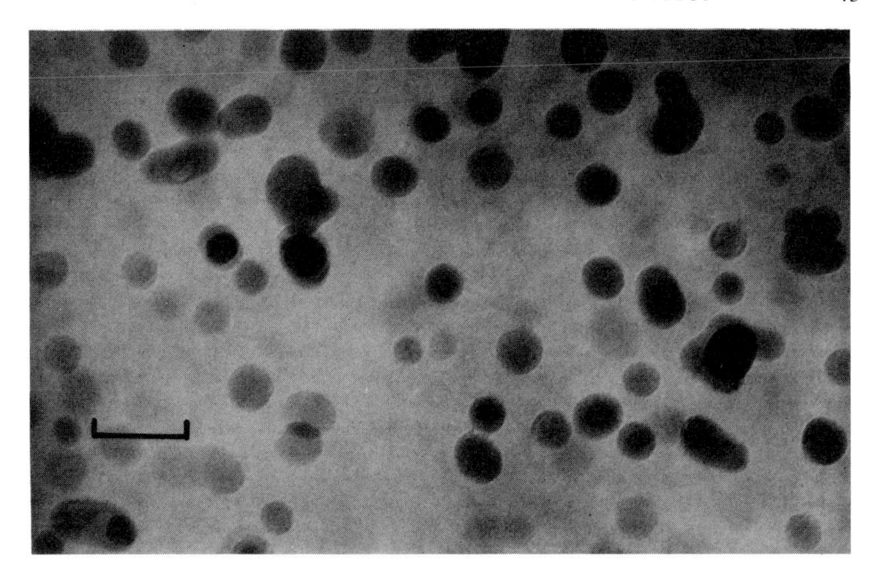

FIG. 14. Phase-separated microstructure of 30 Li_2O 70 SiO_2 glass (bar ≡ 1000 Å).

traversed. For glass-ceramics employing normal cooling rates, therefore, phase separation, if it occurs, is probably more likely to arise by nucleation and growth than by spinodal decomposition.

Figure 14 is an electron micrograph of a glass of the molecular composition $30Li_2O$ $70SiO_2$ that has been heat treated for 1 hr at 550°C. In this case the droplet form of the second phase clearly suggests that a nucleation and growth process could have been responsible for its development. Figure 15 shows the microstructure of a phase-separated borosilicate glass and here the high interconnectivity of both phases would be consistent with spinodal decomposition.

b. Effects of glass-in-glass phase separation upon crystal nucleation and growth processes. Phillips and McMillan (1965) investigated the crystallisation process in glasses of the Li_2O–SiO_2–P_2O_5 type by electron microscopy and electrical conductivity measurements and suggested that the development of fine grained microstructures was significantly enhanced if prior glass phase separation had occurred. They attributed this to an increase of nucleation density and to a reduction of crystal growth rate owing to the presence of SiO_2-rich phase-separated regions which tended to enclose the growing lithium disilicate crystals.

Burnett and Douglas (1971) observed that crystal growth rates in glasses of the Na_2O–BaO–SiO_2 type were reduced if prior phase separation had

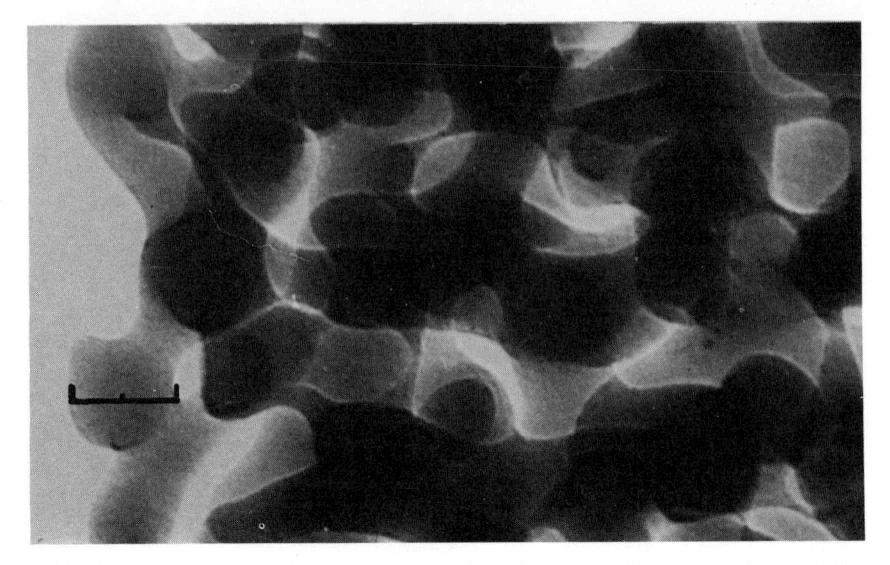

FIG. 15. Phase-separated microstructure of a sodium borosilicate glass (bar ≡ 5000 Å).

occurred. The effect appeared to involve "mechanical interference" of the crystal growth front by the phase-separated particles. These authors were of the opinion, however, that a direct effect of phase separation on nucleation rate did not occur for the system investigated.

In a later study, Harper and McMillan (1972) investigated crystal nucleation and growth processes in glasses of the Li_2O–SiO_2–P_2O_5 types. On

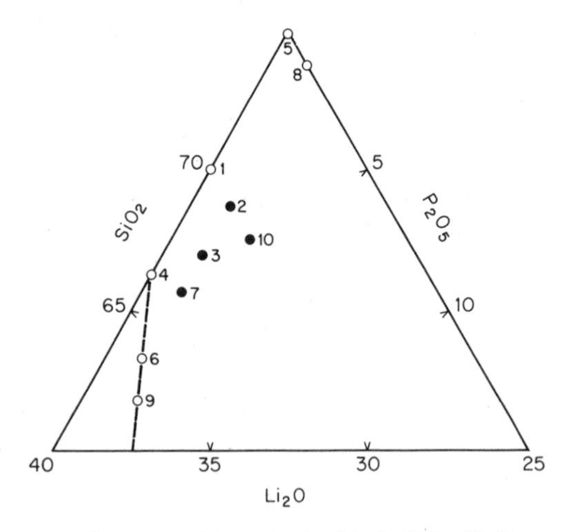

FIG. 16. Glass-ceramic compositions in the Li_2O–SiO_2–P_2O_5 system.

the basis of a thermodynamic model, they predicted that compositions of the general formula $x(Li_2O.2SiO_2)$: $(1-x)$ $(Li_2O.3P_2O_5)$ represented by the broken line in Fig. 16 would be single phase whereas compositions on the silica-rich side of this line should separate into two phases. Experiments confirmed these predictions and enabled studies of crystallisation to be undertaken on both single and two-phase glasses either containing or free from P_2O_5 and having the compositions shown in Fig. 16. The glasses were subjected to a two-stage heat treatment of 500°C/1 hour for nucleation followed by a crystallisation treatment of 750°C/1 hour. Mean crystal diameters in the heat-treated glasses were determined from optical and electron microscopy and are given in Table VI together with the volume fractions of silica-rich droplet phase calculated from the thermodynamic model.

TABLE VI

MEAN CRYSTAL DIAMETERS AND PREDICTED VOLUME FRACTIONS OF SILICA-RICH PHASE FOR Li_2O–SiO_2–P_2O_5 GLASS-CERAMICS

Composition (mol. %)				Crystal diameter (μm)	Volume fraction of SiO_2-rich phase
No.	SiO_2	Li_2O	P_2O_5		
1	70	30	—	126·5	0·124
2	69	30	1	11·7	0·183
3	67·5	31·5	1	42·8	0·130
4	66·7	33·3	—	340·0	0
5	75	25	—	188·0	0.298
6	64	35	1	65·1	0
7	66	33	1	53·1	0·075
8	74	25	1	0·37	0·353
9	61·3	36·7	2	1.23	0
10	68	30	2	0.18	0.241

Summarising the results: it appears that glasses which contain no P_2O_5 form polycrystalline materials with coarse microstructures (ie the nucleation efficiency is low) and while prior glass-in-glass phase separation leads to some improvement in nucleation density, the effect is not clear cut. Single phase glasses which contain P_2O_5 show increasing nucleation densities as the proportion of P_2O_5 is increased. However, the highest nucleation efficiencies

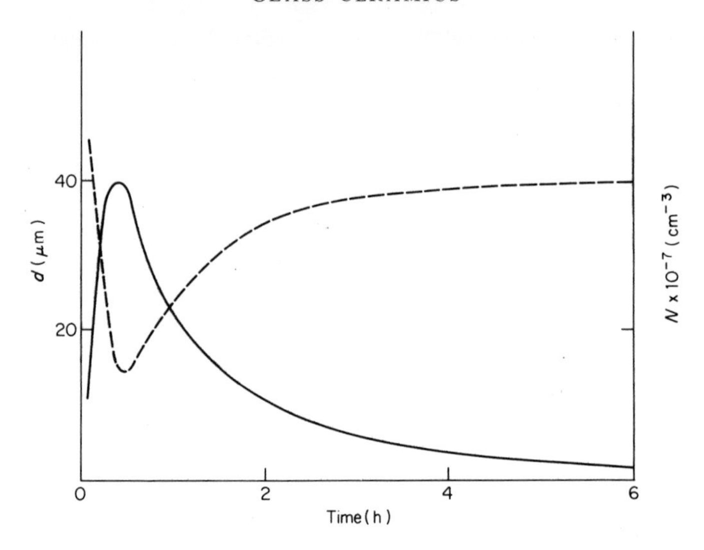

FIG. 17. Effect of duration of nucleation at 550°C on the nucleation density, N, and the mean crystal diameter, d, for a glass of the molar composition Li_2O 30; SiO_2 69; P_2O_5 1 (d,---; N,——).

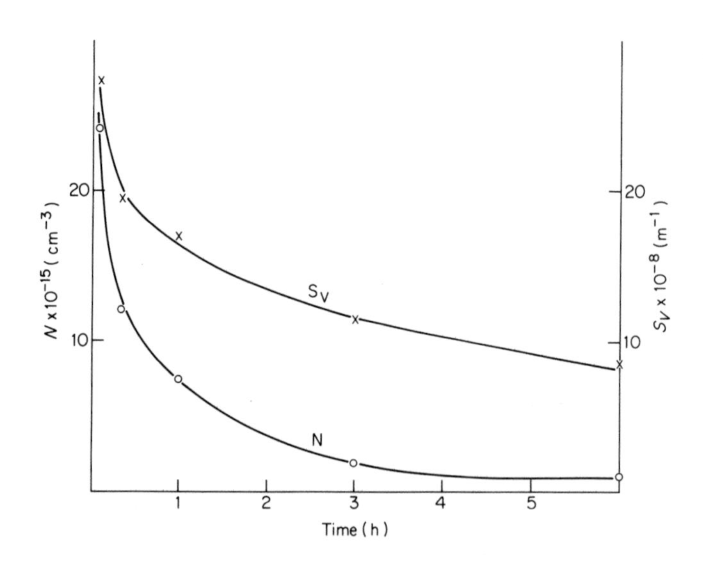

FIG. 18. Influence of heat treatment at 550°C upon number density, N, and interfacial area, S_v, of phase-separated particles in a glass of the molar composition Li_2O 30; SiO_2 69; P_2O_5 1.

are obtained for glasses which contain P_2O_5 and which exhibit prior phase separation.

To investigate whether there was a direct connection between nucleation efficiency and parameters describing the morphology of phase separation, such as number of phase-separated particles per unit volume or interfacial area, McMillan (1974) studied the crystallisation of a glass of the molar composition Li_2O 30, SiO_2 69, P_2O_5 1. Specimens were given varying nucleation treatments between 0·1 and 6 hours at 550°C followed by a standard crystallisation treatment of 1 hour at 750°C. Figure 17 shows the variation in nucleation density, N, and mean crystal diameter, d. Maximum nucleation efficiency was achieved with a duration of 20 minutes at 550°C. There was no simple correlation between this observation and the phase-separated microstructure since as Fig. 18 shows, the number density of phase-separated particles and their interfacial area continued to decrease throughout the heat-treatment period. This result does not support the idea that the phase-separated particles provide nucleation sites for subsequent crystallisation.

There is little doubt, however, from the earlier work that the occurrence of prior separation does enhance crystal nucleation rates. To reconcile these apparently contradictory findings McMillan (1974) proposed that during the nucleation treatment, crystal nuclei arise in the matrix phase of the two-phase glass, initially in increasing numbers, but ultimately the number of nuclei decreases because of a coarsening process.

The occurrence of phase separation in the glass could increase the activation energy of diffusion and thereby hinder the coarsening process. Thus phase separation may act to conserve nuclei so that during the subsequent crystal growth stage, growth will occur from a larger number of sites.

The beneficial effect of prior phase separation on nucleation density is therefore seen as an indirect one, rather than the direct provision of nucleation sites.

Harper and McMillan (1972) used measurements of dielectric loss tangent on glasses of the Li_2O–SiO_2–P_2O_5 type to investigate crystal growth rates at a temperature of 600°C. Their results indicated that prior phase separation led to a reduction of crystal growth rate possibly by a "mechanical interference" effect.

On the basis of these investigations there is little doubt that prior glass-in-glass phase separation has profound effects upon the microstructures developed by crystallisation of glasses. The enhancement of the effective nucleation density combined with the reduction of crystal growth rate are factors that both favour the development of fine-grained microstructures and this is desirable for glass-ceramics production.

3. *Devitrification of Glass*

a. General factors. Devitrification implies the growth of crystalline material in the glass and in normal glass-making practice steps are taken to prevent its occurrence. If devitrification occurs during the shaping of the glass it has a very harmful effect since it can lead to sudden and unpredictable changes of viscosity which interfere with the working of glass-shaping machinery. Also the growth of crystalline phases which may differ markedly in thermal expansion coefficient from the parent glass can lead to the generation of dangerously high stresses in the glass due to the differential contraction which occurs on cooling. Devitrification can occur due to the selection of an unsuitable glass composition, although this would be unlikely since the compositions of commercial glasses are carefully chosen so as to be resistant to devitrification. If, however, the composition has been changed locally due to prolonged contact and reaction with the furnace refractories in stagnant regions of the melting furnace, devitrification may occur. Glass-shaping processes which entail holding the glass for long periods within the "critical" zone of temperatures where crystal growth could occur are to be avoided. For some glasses, which are not particularly stable towards devitrification, there may be limitations with regard to the shaping processes which can be used and only those processes which permit rapid cooling of the glass through the critical zone of temperature will be applicable. This situation can exist for some of the special glass compositions used for glass-ceramic production.

Although the basis of the glass-ceramic process is to accomplish devitrification of glasses, this process must be a controlled one to ensure that the final material has the correct microstructure and properties. Premature devitrification, occurring during the shaping of the glass, is highly undesirable and must be avoided. Crystal growth at this stage is unlikely to result in the very fine-grained structure which is required but is more likely to give rise to a coarse structure consisting of a relatively few large crystals. Such a material would be mechanically weak and would be of little practical value.

Devitrification, or rather its avoidance, is a subject of great importance to the glass-maker but systematic studies have been relatively rare until recent years, partly owing to the experimental difficulties. A further difficulty is that the crystal phase that separates out is not necessarily the most stable phase at the temperature of observation. For example, silica often cyrstallises out as cristobalite under conditions where tridymite is the stable form.

In the devitrification of ordinary glasses, the surface plays an important role since crystallisation proceeds much more rapidly than within the bulk and in most cases crystallisation is initiated at the surface. There may be a number of reasons for this, including a difference of chemical composition between surface and interior regions. Loss of volatile constituents from the glass surface during the shaping operation could account for such a difference and also the fact that constituents which lower the surface energy will tend to concentrate in the surface. It seems probable, however, that the most significant factor is the existence of heterogeneities at the surface which can promote crystal nucleation there. These heterogeneities may be dust particles or minute scratches or flaws. Ernsberger (1962) showed that scratches could act in this way and the work of Partridge and McMillan (1974) on the surface crystallisation of $ZnO–Al_2O_3–SiO_2$ glasses also supported the idea that flaws in the surface led to enhanced crystallisation. The presence of gas bubbles or foreign particles within the interior of the glass could also lead to more rapid devitrification in the surrounding regions. Gross inhomogeneity of the glass which resulted in regions having unstable compositions would also affect the course of devitrification.

From the foregoing it will be clear that devitrification which can occur during the reheating of normal glasses is largely an uncontrolled process depending upon chance effects such as the presence, in particular regions of the surface, of foreign particles which can act as nucleation catalysts. Almost always, glass devitrified in this way shows a characteristic structure after completion of the crystallisation process. The crystals grow inwards from the glass surface in the form of needles and quite often they are oriented at 90° to the glass surface so that the two "fronts" of the crystal growth meet in the centre of the glass. The crystals are generally large and the orientation effect is such that the material will be mechanically weak.

b. Experimental investigations. Mention has already been made (chapter 1) of the work of Réaumur (1739) who produced porcelain-like articles by heat-treating soda-lime-silica glass. In this process, crystallisation was initiated by surface nucleation and for the reasons given above, the articles would be rather weak and brittle. An attempt to make commercial use of crystallised soda-lime-silica glass was undertaken by Garchy in France in about 1896 who produced an artificial stone by the heat-treatment of pressed blocks prepared from scrap glass. Pavements of Garchy stone were laid in certain Paris streets and these were sufficiently durable to survive for many years.

In fundamental studies of devitrification a number of investigators have examined the influence of the degree of supercooling upon the linear growth rate of crystals. This work showed that the growth rate, which is of course zero at the liquidus temperature, rises to a maximum at some temperature below the liquidus and then falls again. The type of growth rate curve usually obtained is illustrated in Fig. 19 which depicts the results obtained by Swift (1947) for devitrite crystallising from a soda-lime-silica glass. If the activation energy for crystal growth is controlled by a diffusion process, the maximum rate of crystal growth should be proportional to the fluidity of the glass, which

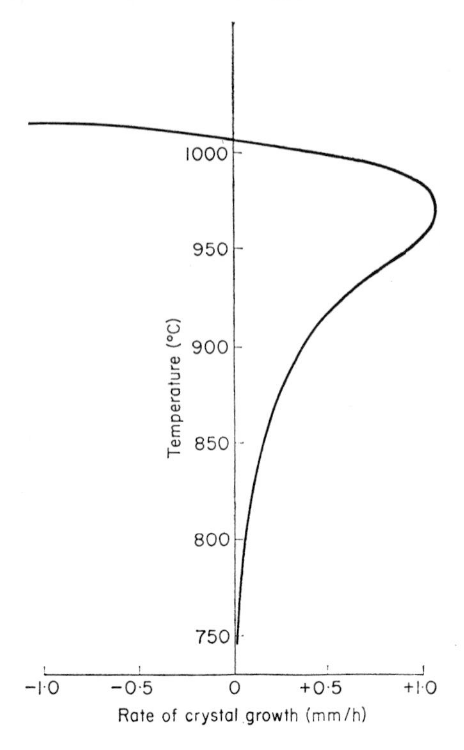

FIG. 19. Effect of temperature on rate of crystal growth in a soda-lime-silica glass (after H. R. Swift; reproduced by courtesy of the American Ceramic Society).

of course is determined by the temperature. This has been shown to be approximately correct by Littleton (1931) for a series of soda-lime-silica glasses for which the phase crystallising out was tridymite. Preston (1940) showed that below the temperature of maximum growth a relationship existed between the logarithm of the growth rate of devitrite crystals and the inverse of the absolute temperature and suggested an empirical expression for growth rate I of the form:

$$I = K(T_L - T)/\eta \qquad (30)$$

where T_L is the liquidus temperature, T is the temperature at which crystal growth rates are measured, η is the coefficient of viscosity, and K is a constant.

This empirical expression is, of course, identical with Eqn (26) discussed on p. 38.

Stanworth (1950) pointed out that the activation energy of the crystallisation process for the glass studied by Preston is much closer to that for electrical conductivity and is markedly lower than the activation energy for viscous flow. A similar observation was made by Cox and Kirby (1947) who studied the crystallisation of cristobalite in a borosilicate glass. These results mean that the activation energy controlling the devitrification process approximates to that for the diffusion of alkali metal ions (sodium) through the glass since the latter process determines the electrical conductivity.

The work of H. R. Swift also illustrates another interesting feature, namely that crystals can be superheated above the liquidus temperature and that a region of "negative growth" exists which is continuous with the growth rate curve below the liquidus. In experiments in which the CaO content was partly replaced by MgO, Swift showed that in amounts up to about 6 per cent this oxide caused a reduction of the liquidus temperatures and growth rates for both cristobalite and devitrite which appeared as crystalline phases in the based glass. This fact is of practical importance since reduction of liquidus temperature improves the stability of a glass against devitrification. It is therefore clear why commercial soda-lime-silica glasses contain a proportion of MgO in their compositions. If the MgO content was increased above 6 per cent, however, diopside ($CaO.MgO.SiO_2$) and sodium magnesium silicate ($Na_2O.2MgO.SiO_2$) appeared as new phases and the stability of the glass was thereby impaired. Inclusion of alumina in the base glass also reduced the crystal growth rate for initial additions, but again at high concentrations, the glass stability decreased owing to the precipitation of wollastonite ($CaO.SiO_2$). In all of the experiments, crystal growth was from the surface.

The complexity of the devitrification process in multicomponent glasses was further emphasised in the studies of McMillan *et al.* (1968a) on relatively simple compositions of the $ZnO–Al_2O_3–SiO_2$ type. The results of crystal growth rate studies using a hot-stage microscope for a glass of the weight percentage composition: ZnO 44·7, Al_2O_3 14·0, SiO_2 41·3 are summarised in Fig. 20. Four crystal types developed, as shown. The initial phase which developed between 800 and 950°C had a modified keatite structure and this transformed irreversibly into zinc orthosilicate at about 980°C. At higher temperatures, zinc aluminate and cristobalite crystals developed.

Modification of the base glass composition by the inclusion of alkali and alkaline earth oxides, B_2O_3, TiO_2, ZrO_2 or P_2O_5 in amounts ranging from 1 to 3 per cent resulted in changes in the crystallisation behaviour. Inclusion of Li_2O had a particularly striking effect since the growth rate of the keatite phase at 800°C was increased from 0·067 μm sec^{-1} to 0·75 μm sec^{-1} by 1 per cent addition and increase of the Li_2O content to 3 per cent resulted in a growth rate of 5·9 μm sec^{-1}. These large changes were attributed mainly to the effect of Li_2O in reducing the viscosity of the glass at 800°C from $5{\cdot}4 \times 10^7$ poise for the base composition to $3{\cdot}2 \times 10^4$ poise for the glass containing 3 per cent Li_2O. Interestingly, 1 per cent additions of Na_2O or K_2O had a rather different effect in that they suppressed the formation of the keatite phase and

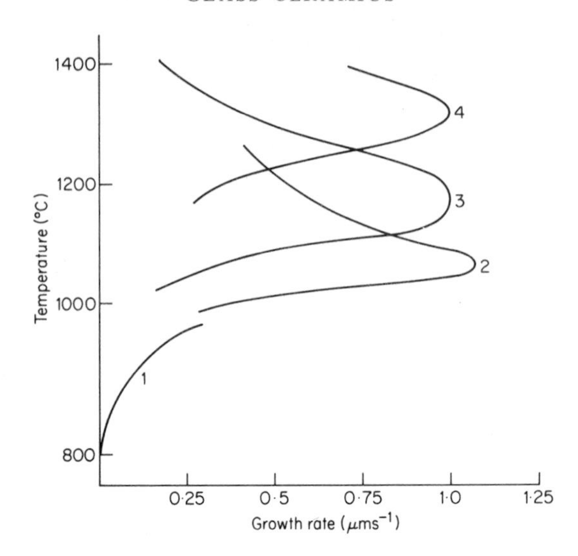

FIG. 20. Crystal growth curves versus temperature curves for a glass of the weight percentage composition: ZnO 44·7; Al_2O_3 14·0; SiO_2 41·3.

resulted in the formation of albite ($Na.Al.Si_3O_8$) and willemite ($Zn_2.SiO_4$) respectively as the initial phases. Other oxide additions influenced the crystal growth rate only slightly and it may be significant that their effects on the glass viscosity were also comparatively small. It is worth noting, however, that two oxides led to a reduced growth rate for the keatite phase, namely MgO and CaO in 1 per cent concentrations.

A further point that emerged from this study concerned the nature of the modified keatite phase. Since this transformed into zinc orthosilicate, it was deduced that it had the same chemical composition as the latter phase namely $2ZnO.SiO_2$. On the basis of this, the suggestion was made that in the modified keatite phase zinc ions were present both in tetrahedral coordination replacing silicon ions and also in octahedral sites to ensure overall charge neutrality in the crystal structure.

Matusita *et al.* (1975) investigated the devitrification of glasses of the general molar composition: 25 Li_2O, 75 SiO_2, 3 RO_n where RO_n representing a wide range of alkali and alkaline earth oxides as well as those of network-forming cations. Sub-liquidus immiscibility was observed for the glasses and it was noted that while the miscibility temperature increased with increasing ionic field strength, the results for network modifying cations fell on a different curve from those for network formers. The glass containing P_2O_5 as the additive, had the highest miscibility temperature. They concluded that the occurrence of immiscibility significantly influenced both nucleation and growth rates and suggested that enhanced nucleation might occur near the interface between the two glassy phases.

McMillan and Phillips (1962) investigated the effects of small compositional changes on crystal growth rates in melts of the general molar composition 35 Li_2O $(65-x)$ SiO_2 xRO_n where RO_n represented Al_2O_3, TiO_2, V_2O_5 or P_2O_5. The crystal phase appearing in the temperature range investigated (~ 800 to ~ 1100°C) was lithium disilicate.

For the base composition, with no additive, it was found that the experimental results gave a reasonable fit to the equation of Hillig and Turnbull (1956)

$$u = A\Delta T^{1\cdot 75}/\eta \tag{31}$$

where u is the growth rate, η is the coefficient of viscosity, ΔT is the undercooling and A is a constant.

Addition of the various oxides in 1 per cent concentrations, resulted in profound changes in the crystal growth rates as shown in Fig. 21. The effect of P_2O_5 in reducing the crystal growth rate was particularly striking while that of V_2O_5 was also significant. Measurements showed that the viscosity

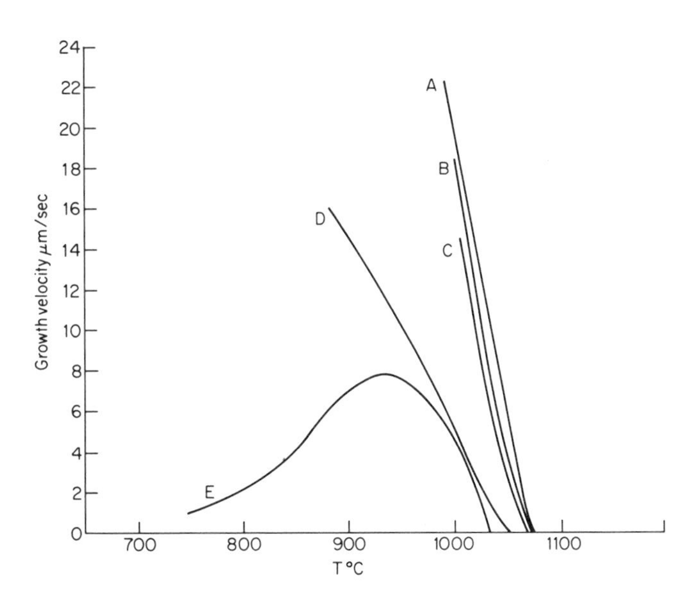

FIG. 21. Crystal growth-temperature curves for Li_2O–SiO_2 glasses. Molecular percentage compositions:
A Li_2O 35; SiO_2 65
B Li_2O 35; SiO_2 64; Al_2O_3 1
C Li_2O 35; SiO_2 64; TiO_2 1
D Li_2O 35; SiO_2 64; V_2O_5 1
E Li_2O 35; SiO_2 64; P_2O_5 1

characteristics of the glass were not significantly changed by the inclusion of P_2O_5 and neither was the electrical conductivity. On this basis it was felt that explanations based on changes in the activation energies of viscous flow or lithium ion diffusion were not tenable. It was suggested that phosphorus and vanadium ions might be incorporated in the lattice of the growing crystal, thereby affecting the activation energy of crystal growth.

Another example of an "impurity" affecting crystal growth was shown by the work of Hibberd and McMillan (1975) who studied the effects of small concentrations of water, present as hydroxyl ions, on the crystallisation behaviour of a glass of the molar composition Li_2O 35.SiO_2 65. Increase of the hydroxyl ion concentration resulted in a displacement of the maximum in the crystal growth curves significantly towards lower temperatures as shown in Fig. 22. The effect is surprisingly large and is partly due to a reduction of viscosity and activation energy of viscous flow that was shown to occur. It is probably also significant that the activation energy for lithium ion diffusion, as shown by electrical conductivity and dielectric loss measurements, was lowered by increasing hydroxyl ion concentrations. The effect on crystal growth kinetics appeared, however, to be larger than could be explained by these factors alone.

Uhlmann (1972) has discussed the effects of impurities on crystal growth in

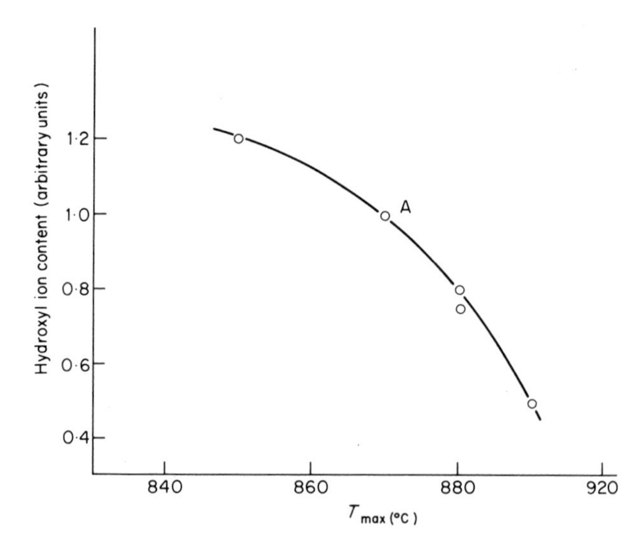

FIG. 22. Effect of hydroxyl ion content on temperature corresponding with maximum growth rate (T_{max}) of lithium disilicate in a glass of the molecular percentage composition Li_2O 35; SiO_2 64; P_2O 1. (The relative OH contents were determined by infrared absorption spectroscopy; for point A the actual water content of the glass was estimated to be 0·03 weight per cent.)

glass-forming systems. He pointed out that while for many materials the presence of impurities causes a reduction of the crystallisation rate, because of diffusional processes required to permit crystallisation or because of poisoning of preferred growth sites, in the case of oxide glasses increased rates of crystal growth may be promoted. Examples of this are given by H_2O, O_2 and Na in the case of silica glass and H_2O in the case of vitreous boric oxide.

In summary, it is clear that compositional changes that result in a reduction of viscosity can result in enhanced crystallisation rates in glasses but there appear to be other effects possibly involving the incorporation of impurities into the growing crystal which affect the activation energy of crystal growth. The effect of small concentrations of water upon crystallisation kinetics is especially interesting since this is present in practically all oxide glasses as an impurity and its possible effects have not generally been taken into account in studies of the devitrification of glass.

c. Crystal growth morphology. An important aspect of the crystallisation of glasses concerns the growth morphology of the crystal phase. Uhlmann (1971) considers that a crucial factor in determining the growth front morphology is the entropy of fusion. Materials having low entrophies of fusion ($\Delta S_{fM} > 2R$) grow with non-faceted interfacial morphologies and the growth rates are nearly isotropic. Materials for which entropies of fusion are large ($\Delta S_{fM} < 4-6R$) exhibit atomically smooth interfaces for close packed planes while the less closely packed planes are rough on an atomic scale. This behaviour is accompanied by large anisotropies of growth rate between the different orientations. Faceting of the most closely packed interfaces is also likely.

For many materials having high entropies of fusion, undergoing crystallisation at large undercoolings, nucleation of new crystals having different orientations can occur either at or in advance of the growth front; this type of nucleation which can result in spherulitic growth is favoured by low molecular mobility (ie high viscosity of the glass phase and by the presence of impurities).

The question of spherulitic growth is of sufficient importance from the point of view of glass-ceramics formation to merit discussion. All spherulites comprise fibrous crystals radiating from a common centre and many, as the name implies, have a spherical geometry but others may have a sheaf-like appearance. Spherulites occur in organic polymers, in the crystallisation of gels, in igneous rocks and in commercial glasses when the glass devitrifies under conditions where its temperature is held constant for a prolonged period, eg in stagnant regions in a glass tank furnace. The occurrence of spherulite formation in glass-ceramics systems is not uncommon although it results in a microstructure that is not very favourable to the achievement of

FIG. 23. Spherulite formation in a glass of the molecular percentage composition Li_2O 30; SiO_2 70. The glass was heat treated for 9 hours at 550°C followed by 1 h at 750°C. (×220)

high mechanical strengths. Figure 23 illustrates spherulite formation in a crystallised Li_2O–SiO_2 glass.

Investigations of spherulitic growth, mainly for polymers, have shown that crystallographic orientation is not maintained in the fibrous structure, in contrast to the situation for dendritic growth, and the branching angles are not related to the crystal lattice.

Keith and Padden (1963) have proposed a theory for spherulitic growth in which they emphasise the importance of the ability of the crystals to branch at a stage subsequent to nucleation. They also regard the presence of impurities in the melt as crucial; these, rejected by the advancing crystal, become concentrated at the interface and diminish still further the already low growth rate. Growth is envisaged as proceeding by fibrous projections separated by regions of uncrystallised glass which eventually crystallises by slow lateral growth. Branching of the fibres occurs by the formation of perturbations on their surfaces and since the critical size of a perturbation to permit growth decreases with increased undercooling, the branched frequency increases.

The kinetics of spherulitic growth can be analysed in terms of the Johnson-Male-Avrami equation:

$$V_f = 1 - \exp(-Kt^n) \tag{32}$$

A value of the exponent, n, of 3 implies a constant radial growth rate with no formation of new nuclei and a value of 4 implies both a constant growth rate and a constant nucleation rate. For polyethylene and natural rubber the value of n lies between 3 and 4 implying an intermediate situation but the results for oxide glasses are less easily understood because values of n between 1 and 2 are found for Li_2O–SiO_2, BaO–SiO_2 and Na_2O–P_2O_5 glasses.

A possible, but improbable, explanation for a value of $n = 2$ is that laminar growth takes place requiring one layer to be completed before the next commences to form. An alternative explanation, which may be consistent with experimental observation, is that the crystallisation process is controlled by the rate of nucleation of new fibres. It is assumed that the growth rate remains constant and the nucleation rate is dependent on t^{-2}. Hence the density of fibres would decrease as the spherulitic growth proceeds, as observed experimentally.

Heat treatment of spherulite forming glasses at higher temperatures than those causing spherulitic growth can result in complete structural modification converting the crystals to lath- or needle-like forms. Thus a more desirable glass-ceramic microstructure can be generated; this phenomenon has been observed for the BaO–SiO_2 system and for the ZnO–Al_2O_3–SiO_2 system.

4. *Requirements for Controlled Devitrification of Glasses*

The aim of the glass-ceramics technologist is to produce polycrystalline ceramic materials by the controlled crystallisation of specially formulated glasses using a heat treatment cycle that is both economical and commercially attainable.

Important objectives include the attainment of a fine grained microstructure that will confer good mechanical properties, the production of crystal phases that will allow other important physical properties to be attained (eg a specified thermal expansion coefficient) and regulation of the volume fraction and composition of the residual glass phase.

Attainment of these objectives depends upon the interaction of a number of complex factors but nevertheless it is possible to lay down general principles that must be followed.

First, it is clear that the achievement of the desired microstructure will be strongly dependent on inducing a high crystal nucleation rate within the glass. As we have seen, surface nucleation of a bulk glass does not generally lead to a crystallised product having satisfactory physical properties. This statement must be qualified, however, by saying that certain surface-crystallised glasses can be produced having excellent mechanical properties and the techniques employed will be described in chapter 4. Also, if crystallisable glasses are reduced to a fine powder and then fabricated into bulk materials by a sintering process, the crystal nuclei generated on the surfaces of the glass particles occur in sufficient numbers to enable a satisfactory microstructure to be obtained.

For the achievement of bulk crystallisation, it is necessary to ensure that a high density of crystal nuclei arises within the body of the material. In general, for the majority of glasses this cannot be achieved by homogeneous nucleation but there are undoubtedly certain restricted ranges of composition (those with high lithia contents for example) where homogeneous nucleation may occur to a sufficient extent to allow glass-ceramics to be made.

The optimum technique for producing a fine-grained glass-ceramic is to generate crystal nuclei in the glass at temperatures below those at which major crystalline phases could grow at a significant rate. High densities of nucleating particles and uniform dispersion are essential to promote subsequent crystal growth from many closely spaced centres. Even with random orientation of the crystals, a glass-ceramic containing coarse crystals is unlikely to be so satisfactory as one which has a dense fine-grained structure.

The addition of constituents known as nucleating agents to the glass to promote the development of a high density of internal nucleation sites is the key to the achievement of controlled crystallisation. The nucleating agent is soluble in the molten glass but either during controlled cooling or during reheating of the initially chilled glass it takes part in or promotes structural changes with the glass. The nucleating agent may, in some cases, precipitate out by homogeneous nucleation in a highly dispersed form. The nuclei formed in this way can then promote heterogeneous nucleation of major crystal phases.

For nucleating agents to function in this manner, certain basic requirements must be fulfilled. The rate of homogeneous nucleation must be high, requiring a low free energy of activation for nucleation. This in turn requires a high degree of supersaturation and a low interfacial energy between the nuclei and the glass. In addition the activation energy for diffusion of atoms or ions which makes up the nuclei must be low, to allow ready growth of embryos to the stable size.

The potency of the nucleating particles in promoting growth of major phases will depend on factors such as the interfacial energy between the nucleating particle and the phase being crystallised and the degree of similarity between the lattice parameters of the nucleus and the precipitated phase.

In addition to the possibility that nucleating agents may serve to provide crystalline nuclei directly as discussed above, there is another possibility. This is that the nucleating agent promotes sub-liquidus phase separation and thereby favourably influence the kinetics of crystal nucleation. Stookey (1959a) discussed this possibility and pointed out that two-phase liquid (or glass) separation is more likely to occur than homogeneous nucleation of crystallisation. The interfacial energy between two liquids is very small and may be almost zero whereas that between a crystalline nucleus and a glass is appreciable.

Two-phase glass separation occurring prior to crystal nucleation and growth could influence the behaviour of the glass during reheating in several ways. As Uhlmann (1970) has pointed out, a larger driving force for nucleation may arise than existed previously, interfaces that could act as preferred sites for nucleation are created, atomic mobility may be enhanced and the interfacial regions between the two phases may be enriched in some component providing, locally, a higher driving force for nucleation or higher mobility. Uhlmann was of the opinion that an increase in atomic mobility was the most important of these factors.

The changes in structure and mobility of various atomic species that result from glass phase-separation may increase the probability of homogeneous nucleation within one or both of the two phases and nuclei arising in this way may serve to provide heterogeneous nucleation of other crystal phases.

In addition to the various mechanisms already discussed by which additives to glass may influence the crystallisation process there is another possible effect which must be mentioned. As we have seen, the interfacial energy between the precipitated phase and the mother phase constitutes a barrier to nucleation, especially when the precipitated phase is crystalline. If this interfacial energy could be lowered by some method, nucleation would be facilitated. Lowering of the surface free energy might occur if a "surface-active" constituent became concentrated at the solid-liquid interface. Hillig (1962) has discussed this possibility and suggested that a lowering of surface free energy would result from a diffuse boundary between the two phases and also that ions (such as the titanium ion) which increase the polarizability of oxygen ions would lower the work associated with creating a melt-solid interface. Such effects as these may be difficult to check experimentally but nervertheless they cannot be discounted in any considerations of catalysed crystallisation.

Finally, the role of crystal growth rate in determining the microstructure of a crystallised glass should be mentioned. While growth must proceed at a rate that is compatible with crystallising the bulk of the glass in an acceptable and economic time, the occurrence of very high growth rates is unlikely to promote the development of a fine-grained microstructure. Under such conditions, once growth is initiated at a nucleation site, the crystal growth front could move so rapidly as to prevent initiation of growth at nearly potential sites. Also rapid growth implies that conditions for coarsening of the microstructure exist in which small crystals redissolve and large crystals are enlarged. In this process the overall interfacial area between the crystal and glass phase is reduced thereby lowering the surface free energy of the composite material.

On this basis, it appears that the choice of glass compositions to achieve moderate growth rates and the inclusion of constituents that restrain crystal growth are also important factors in the achievement of the desired control of the crystallisation process.

The foregoing discussion has been concerned chiefly with methods of controlling the size and numbers of crystals produced, which is extremely important. In addition to this, however, the proper selection of the nucleation catalyst can also enable a certain degree of control to be excercised over the types of crystals formed. A multicomponent glass may contain the constituents of two or more crystalline phases in solution, some or all of which may be caused to precipitate under the correct conditions. If the nucleation catalyst has a structure similar to one phase, this phase will probably precipitate most readily even though it may not be the equilibrium phase for the particular conditions of heat treatment. With other catalysts, of course, the equilibrium phase may crystallise out. The changes of composition of the residual glass which accompany the first development of a crystal phase may cause the precipitation of still further phases. Thus the choice of the nucleation catalyst can sometimes exert a marked influence upon the crystal phases present in the glass-ceramics. Even with the same crystallisation catalyst, changes of heat treatment process can often alter the nature of the crystalline phases finally present.

Chapter 3

NUCLEATING AGENTS AND GLASS TYPES FOR CONTROLLED CRYSTALLISATION

A. Types of Nucleating Agents

1. *Nucleation by Separation of Metals*

One of the chief requirements for a crystallisation catalyst or nucleating agent is that it must be capable of existing in the glass in the form of a dispersion of particles of colloidal dimensions. There are a number of metals which can exist in this form in glass and it is therefore likely that they could act as nucleating agents to promote the controlled crystallisation of glasses. Suitable metals will be those whose oxides are readily broken down to the metal either by the application of heat alone or by the use of reducing conditions during melting.

a. Metallic colloid formation in glasses. Copper is an example of a metal which can be obtained in the elementary form in glass if the glass is suitably reduced during melting. Silver oxide readily breaks down to give the metal when it is heated, so that elementary silver can be obtained in glass even if oxidising conditions prevail during melting. Similarly, gold and the platinum group metals are readily produced in their metallic states in glass. It is important to understand how these various metals can be obtained in the form of colloidal dispersions in glasses and we shall therefore examine the processes involved in some detail.

Weyl (1951) considers that metals such as silver, gold or copper can be obtained in true solution in glass melts. That is, the metal is present in a state of atomic subdivision. Suitable heat-treatment of the glass either by controlled cooling of the molten glass or by reheating of the chilled glass leads first to the production of tiny aggregates of metal atoms. These act as nuclei for a subsequent crystal growth process by diffusion of metal atoms which become attached to the surface of the nuclei. The particles grow to colloidal dimensions and in extreme cases they may become visible as actual crystals. Tin oxide (usually stannous oxide) plays an important role in the formation of metallic dispersions in glasses and its function may partly be to increase the solubility of the metals in the glass melt although in certain cases, notably for glasses containing copper, it undoubtedly acts as a reducing agent.

Copper can produce a number of colouring or optical effects in glass though here we are chiefly concerned with only one of these, the formation of the copper ruby glass which contains metallic copper in the form of crystals of colloidal dimensions.

If a glass containing copper is melted under oxidising conditions the copper is mainly present as cupric Cu^{2+} ions which give rise to a blue colour but cuprous Cu^{+} ions will also be present; the latter do not cause absorption in the visible region of the spectrum. With increased reduction, the equilibrium will be shifted causing an increase in the proportion of cuprous ions and ultimately elemental copper will be formed. The state of the copper in the glass is thus strongly dependent on the melting conditions and also on the composition of the glass. Stannous oxide, SnO appears to be a particularly suitable reducing agent in accordance with

$$Sn^{2+} + 2Cu^{+} = Sn^{4+} + 2Cu.$$

A further effect of tin oxide, as mentioned earlier, may be to increase the solubility of metallic copper in the glass and thus enable it to exist in the melt in a state of atomic subdivision. In this condition, copper will not result in visible absorption. Thus if a glass melt containing cuprous ions and elemental copper is cooled moderately rapidly to "freeze in" the high temperature state of the copper, a colourless glass results. If the glass is now reheated to a sufficiently high temperature, a phenomenon known as "striking" takes place in which the deep red colour of the copper ruby develops. It seems likely that this may involve more than one process. Further reduction of Cu^{+} ions to the copper metal takes by the reaction:

$$Cu^{+} + e \rightarrow Cu$$

in which the electrons are donated by Sn^{2+} or Sb^{3+} ions which are thereby oxidised to Sn^{4+} or Sb^{5+} ions. The copper atoms so formed together with those already present are able to undergo diffusion at the temperature of striking and thus by a process of nucleation and growth, copper particles of a sufficient size to affect the transmission of light through the glass are developed.

For the production of copper ruby glasses, high lead glasses are often used, perhaps because metallic copper is particularly soluble in these, though soda-lime-silica compositions can also be employed. The copper is usually present in a concentration of 0.2 to 1 per cent to ensure satisfactory colouration.

When colloidal dispersions of copper are used in glasses to allow controlled devitrification, where colour is of minor importance, smaller concentrations can be employed. For example, concentrations as low as 0·05 per cent have been used successfully to nucleate the crystallisation of Li_2O–AL_2O_3–SiO_2 and Li_2O–ZnO–SiO_2 glasses. It is probable that in these, the colloidal particles of copper could be of a size too small to give a colouring effect but still be of sufficient dimensions to act as heterogeneous nuclei for controlled crystallisation.

The use of metallic silver in the form of a colloidal dispersion as a glass colourant has long been known. Its chief application has been in the staining of glass to produce yellow or amber colours. Silver in its ionic state is far less stable in glass than is copper so that for silicate glasses containing more than 1 per cent of silver, precipitation of the metal occurs even if melting takes place under oxidising conditions. In certain phosphate glasses, however, 10 to 15 per cent of silver ions can be kept in solution.

It seems likely that when a silver-containing glass is melted under oxidising conditions a high proportion of the silver is present as Ag^+ ions but that silver atoms are also present. The Ag^+ ions take part in the glass structure as network modifiers occupying structural sites similar to those occupied by sodium ions. Silver in the form of ions contributes no colour. The silver atoms are also colourless but their presence can be demonstrated by the strong fluorescence which occurs when the glass is irradiated by ultraviolet light.

The equilibrium between Ag^+ ions and Ag atoms is determined by factors such as: the presence of oxidising or reducing agents in the glass batch or the nature of the furnace atmosphere; the melting temperature, since increased temperatures favour reduction to the metallic state; the composition of the base glass which, in addition, influences the solubility of metallic silver.

If a silver-containing glass is cooled it can be obtained in a nearly colourless condition but when it is reheated it "strikes" to give a yellowish amber colour. The formation of the colour is due to the aggregation of silver atoms to form nuclei followed by growth of these nuclei to a size where they are capable of causing selective absorption of light passing through the glass (Weyl, 1951). Slightly reducing conditions are desirable during the melting of silver-containing glasses especially if the base glass is of the soda-lime-silica type. Tin oxide is again a useful constituent since it appears to increase the solubility of metallic silver in the glass at high temperatures and it may also lead to reduction of Ag^+ ions to silver atoms during striking. Formation of nuclei is enhanced by irradiation with ultraviolet light since Badger and Hummel (1945) showed that glasses containing silver strike at a lower temperature if they are irradiated with ultraviolet light before being heat-treated. This phenomenon is of importance in connection with the production of photosensitive glasses and glass-ceramics.

Gold is also well known as a glass colourant and is used to produce particularly beautiful ruby glasses. There is little doubt that the colouring effect is due to the presence of a dispersion of gold crystals of colloidal dimensions. The gold is usually introduced into the glass as gold chloride and, according to Weyl (1951), this breaks down to produce largely gold atoms which are soluble in the glass at high temperatures. At the same time, however, some auric ions, Au^{3+}, may be present in the glass. High lead glasses favour the formation of the gold colour and as little as 1 part in 100 000 parts of glass will give a faint pink colour, whereas deep red colours are obtainable with 1 part of gold in 1000 parts of glass. Gold is much less soluble in lead-free glasses and tin oxide must be present in order to obtain equivalent effects. Antimony oxide also has a beneficial effect on the formation of the colloidal gold dispersion and this point will be discussed in further detail later.

The sequence of events in the formation of gold ruby glass is as follows. At high temperatures the gold is present as an equilibrium mixture of gold atoms, Au, and auric ions, Au^{3+}. If the glass is cooled quickly, the high temperature state of the gold is "frozen in" and the glass is colourless. If the glass is reheated above its annealing point, formation of gold nuclei takes place due to aggregation of gold atoms and growth of these occurs until they achieve sizes of 50 to 500 Å, when they are capable of selectively absorbing light which passes through the glass. It is likely that this does not represent the complete picture because reduction of auric ions to gold atoms may also occur during the reheating process by a reaction which can be represented by:

$$Au^{3+} + 3e \rightarrow Au.$$

The electrons necessary for this reaction can arise by changes in the state of valency of antimony or tin ions present in the glasses:

$$Sb^{3+} \rightarrow Sb^{5+} + 2e$$

and

$$Sn^{2+} \rightarrow Sn^{4+} + 2e.$$

Platinum in the form of a colloidal dispersion in glass has not been used as a colourant because it gives only a greyish tint. However, colloidal dispersions of this metal are of interest in the present connection since the metallic crystals can act as nucleation sites for other supersaturated phases in the glass so that by this means the entire glass can be crystallised.

The process of formation of a colloidal platinum dispersion differs from those for copper, silver or gold. The main difference is that a reheating step is not required. According to Rindone (1962) the platinic chloride, which is the usual source of the metal included in the glass batch, breaks down under the influence of heat and colloidal platinum is formed in the melt.

The presence of ionic forms of platinum at least in some glass compositions cannot be completely ruled out, however. The evidence from this comes from the observation that platinum is slightly soluble in some glasses. While the solubility is vanishingly small if melting is carried out in an inert atmosphere or under vacuum, melting in air or other oxidising atmosphere results in measurable, albeit small, solubility. Thus while platinum metal may be virtually insoluble in the molten glass, the formation of ionic platinum by reaction with oxygen clearly results in solubility.

If, as seems probable, ionic forms of platinum are present in the glass melt, these could undergo subsequent reduction during the cooling of the melt allowing either growth of existing platinum particles or the precipitation of additional particles by a nucleation and growth process.

It is likely that the four metals we have considered would fall into a series:

$$\text{copper} \rightarrow \text{silver} \rightarrow \text{gold} \rightarrow \text{platinum.}$$

In the case of copper, at one extreme, the formation of the colloidal dispersion may be predominantly determined by a reduction effect during the striking of the colour of the type:

$$Cu^+ + e \rightarrow Cu.$$

In this case, the presence of suitable ions of variable valency such as tin, antimony or arsenic plays a vital role in the colloid formation. At the same time copper atoms may be present in the chilled glass and these may also take part in the diffusion process leading to colloid formation. In the case of platinum, at the other extreme, reduction and colloid formation are achieved by the action of heat alone and there is no necessity for a reduction process at low temperature corresponding with the "striking" of the copper glass. Similar behaviour is to be expected for other metals in the platinum group. In the case of silver, the reduction process plays a part in the colloid formation and antimony or tin oxides are still vital. For gold, the ion to metal reduction process, while it still occurs during reheating, probably plays a less important role in colloid formation than for silver or copper.

b. Metallic colloids as nucleation catalysts. Examples of the use of the platinum group metals as nucleation catalysts are given in British Patent no. 863 569 (1961). The metals ruthenium, rhenium, palladium, osmium, iridium and platinum are used in amounts of 0·001 to 0·10 per cent to catalyse the crystallisation of two basic types of glass. The first group of glasses are of the R_2O–BaO–SiO_2 system and have weight percentage compositions in the range: SiO_2: 50–85; BaO: 3–45; Li_2O: 0–25; Na_2O: 0–10 and K_2O: 0–10. Glasses of the second type are based on the Li_2O–Al_2O_3–SiO_2 system and have weight percentage compositions in the range: SiO_2: 60–85; Al_2O_3: 2–25; Li_2O: 6–15 and Na_2O/K_2O: 0–4. During the melting of the glasses the platinum group metal compound in the glass batch decomposes, even under oxidising conditions, and the metal is obtained in the form of a uniform dispersion in the glass. Such glasses can be almost completely crystallised by heat treating them first at a temperature of 580 to 650°C and then by heating to a higher temperature in the range 700 to 850°C. The glass-ceramics produced in this way have good mechanical and electrical properties.

The use of copper, silver or gold either alone or combined with a platinum group metal is described in British Patent no. 863 570 (1961). These metals are used to catalyse the crystallisation of glasses of the two basic types described in the previous paragraph and the heat treatment processes for the production of the glass-ceramics are also similar to those of the earlier example.

Rindone (1962) has made a careful study of the use of platinum as a crystallisation catalyst in a lithium silicate glass of the molecular composition $Li_2O.4SiO_2$. He found that small amounts of platinum enhanced the crystallisation of the glass since not only was the rate of crystallisation increased but a very uniform distribution of small crystals (less than 500 Å in size) could be obtained. He investigated the relationship between platinum concentration and rates of crystallisation of the glass at 600° and 650°C by determining the amount of lithium disilicate crystallising from the glass by X-ray diffraction analysis. It was demonstrated that for both temperatures, optimum concentrations of platinum exist to give maximum crystallisation in a given time. At 600°C the optimum concentration was in the range 0·004 to 0·007 per cent, and at 650°C a concentration of 0·008 to 0·011 per cent Pt gave maximum crystallisation. Amounts of platinum in the region of 0·025 per cent led to a considerable decrease in the amount of lithium disilicate crystallising, probably because under these conditions a form of crystalline silica was precipitated which resembled the high temperature modification of quartz. It was shown that the activation energy of crystallisation of lithium disilicate decreased sharply to a minimum value as the concentration of platinum was increased from zero to 0·005 per cent. For further increases of platinum content the activation energy tended to level off at a slightly higher value. Rindone suggested that the ability of platinum to nucleate lithium disilicate may be related with the fact that the 111 plane of the platinum crystal resembles the 002 plane of $Li_2O.2SiO_2$ with a disregistry of approximately 5 per cent. This falls well within the maximum disregistry of 10 to 15 per cent suggested by Turnbull and Vonnegut (1952) for good nucleation catalysis.

In the same study Rindone made deductions concerning the size of the platinum nuclei and came to the conclusion that for the optimum concentration of 0·005 per cent platinum a nucleus size of 50 Å would provide an ample number of sites for crystallisation and that in this case 10^{13} platinum nuclei would be present in each gramme of glass. This corresponds to an average spacing between particles of about 4500 Å.

Thakur (1971) using a differential thermal analysis technique showed that platinum in a concentration of 8 to 32×10^{-5} mole per 100 gm resulted in bulk nucleation of $Li_2O.2SiO_2$ glass. Silver gave more effective nucleation but higher concentrations ($\sim$6 to 8×10^{-3} mole per 100 gm) were required. Interestingly, it was also shown that elementary silicon acted as a bulk nucleating agent although this was less effective than the metals.

c. Photosensitive reactions in nucleation catalysis. It was mentioned earlier that the striking of the silver colour was assisted by irradiating the glass before heat-treatment. This effect and similar phenomena found with gold and copper formed the basis for the development first of photosensitive glasses and later of photosensitively nucleated glass-ceramics.

The earliest photosensitive glasses contained copper as the active constituent. Dalton (1947) showed that a colourless copper-containing glass, with a composition closely similar to that of a commercial copper ruby glass, after irradiation with ultraviolet light would strike at a lower temperature or in a shorter time than it would normally do so. If the glass were selectively irradiated, as for example by placing a photographic negative on its surface, an image could be developed in the glass when it was suitably heat-treated. The basic mechanism in the photosensitive glass is the formation of a latent image during the irradiation process followed by the "development" of this image during the heat-treatment process. The formation of the latent image consists of the production of copper atoms by a process which can be represented by:

$$Cu^{+} + e \rightarrow Cu.$$

The electrons necessary for this reaction can be derived from:

$$Cu^{+} + h\nu \rightarrow Cu^{2+} + e.$$

The irradiated portions of the glass therefore contain higher concentrations of copper atoms than the unexposed portions and, as a result, nucleation and crystallisation of copper will take place more easily in the irradiated regions.

Stookey (1947) showed that the inclusion of a small amount (up to 0·05 per cent) of cerium dioxide improved the photosensitivity of glasses containing copper as the photosensitive element. The valuable effect of cerium is due to the ease with which electrons are released by the reaction:

$$Ce^{3+} + h\nu \rightarrow Ce^{4+} + e$$

so that for the improved glasses the reaction can be represented by:

$$Ce^{3+} + Cu^{+} + h\nu \rightarrow Ce^{4+} + Cu.$$

In later developments, it was shown that both silver and gold could act as the photosensitive elements in suitable glass compositions and a range of photosensitive glass compositions containing these metals has been described (Stookey, 1950a). The glasses are silicate types containing sodium oxide, aluminium oxide, barium and zinc oxides. The glasses containing gold are melted under oxidising conditions by the inclusion of an alkali metal nitrate in the batch and contain 0·01 to 0·1 per cent of gold which is introduced into the glass batch in the form of a gold chloride solution. The silver-containing glasses are also melted under oxidising conditions and contain silver equivalent to about 0·05 per cent of 0·3 per cent AgCl, this being the form in which the silver is introduced into the glass batch. Glasses in which the sensitive component is copper contain about 0·05 to 1 per cent copper computed as Cu_2O. Such glasses are melted under reducing conditions and the batch therefore contains a reducing agent such as sugar.

The glasses contain up to 0·05 per cent CeO_2 which increases the photosensitivity and also contain up to about 0·1 per cent Sb_2O_3 which has a similar beneficial effect. If such glasses are irradiated by exposure to ultraviolet light and heated to temperatures in the region of 500° to 600°C, the exposed portions of the glass become coloured due to precipitation of the metal in the irradiated portions. It seems likely that the function of the Sb_2O_3, like that of CeO_2, is to donate electrons to the copper, silver or gold ions, converting them to atoms which can aggregate when the glasses are reheated.

In a later development (Stookey, 1950b) photosensitive glasses were produced in which the images developed after selective irradiation and heat-treatment were opaque and consisted of lithium disilicate or barium disilicate crystals. Small amounts of copper, gold or silver were also present and the effect of irradiation with ultraviolet light followed by heat-treatment was first to cause the precipitation of colloidal metal particles in the irradiated areas; these particles then calalysed the crystallisation of the appropriate silicate. This represented an important step in the development of glass-ceramics because it was one of the early examples of the use of controlled (heterogeneous) nucleation to achieve the crystallisation of a material from a glass.

The photosensitive glasses containing lithia are of great interest since some of the first satisfactory glass-ceramics were derived directly from them. The important basic discovery was made that if glasses of this type which have been opacified by precipitation of lithium silicate crystals upon the metallic nuclei are heated in a controlled manner, further crystal growth takes place upon the original tiny crystals until a major proportion of the glass is converted into crystalline compounds. The glasses used (Stookey, 1956) are lithia-alumina-silica types having compositions in the range 60 to 85 per cent SiO_2, 5·5 to 15 per cent Li_2O and 2 to 25 per cent Al_2O_3. In addition the glasses contain a photosensitive metal selected from the group: 0·001 to 0·003 per cent of gold (computed as Au), 0·001 to 0·3 per cent of silver (computed as AgCl) and 0·001 to 1 per cent of copper (computed as Cu_2O). The glasses are first irradiated with ultraviolet light which causes the production of atoms of the photosensitive metal. The next step in the process is to heat the glasses to a temperature between the annealing point ($\log_{10}$ viscosity = 13·4) and the fibre softening point ($\log_{10}$ viscosity = 7·6) of the glass; for the types of glass in question, temperatures in the range 500–540°C are suitable. At this temperature, submicroscopic crystals of copper, silver or gold are first formed by aggregation of metal atoms. These particles then serve to nucleate the crystallisation of a further phase which, in the case of glasses having compositions lying in the range quoted above, is always lithium metasilicate during the initial stages. If the glass is high in lithia and contains K_2O the crystalites may consist entirely of lithium metasilicate ($Li_2O.SiO_2$) but if the silica content of the glass is high, crystallites of lithium disilicate and quartz may form. For glasses high in alumina, crystals of beta-spodumene ($Li_2O.Al_2O_3.4SiO_2$) or of a beta-spodumene-quartz solid solution may develop.

At this stage the glass is merely opacified with tiny crystals of lithium metasilicate or other compounds surrounding the metallic nucleating particles. Between these crystallites there remain regions of glass. If the temperature is raised moderately slowly (5°C per minute) further crystal growth takes place and the crystals formed serve to maintain the rigidity of the material and prevent deformation as the temperature increases. The glass is heated to an upper temperature which does not exceed 950°C and after a suitable holding period of about 1 hour it is cooled. As a result of heat-treatment, the glass is converted into a polycrystalline solid in which the proportion of glassy matrix is quite small. Glass-ceramics produced by this process possess a number of highly desirable properties, including high mechanical strength and good electrical properties.

According to Stookey and Maurer (1962) metal crystals produced by photonucleation provide a catalyst very uniform in size. There is a critical size for these metallic nucleating particles since lithium metasilicate does not begin to crystallise until the metal crystals reach a size of about 80 Å. The factors which determine the critical size of the nucleating particle are uncertain but they may be associated with the smallest stable size of the lithium metasilicate crystal.

Stookey (1959a) has suggested that the effectiveness of metallic nucleation catalysts is related to the similarities between the crystal structures of the metals and of the phase being nucleated. Thus, as he points out, the crystal phases sodium fluoride (NaF), lithium metasilicate (Li_2SiO_3) and barium disilicate ($BaSi_2O_5$), can be nucleated by either gold, silver, copper or platinum and each crystal phase has at least one interatomic distance similar to those in the metal crystals.

It has been shown by Miles and McMillan (1971) that the microstructure of a photosensitively nucleated gold-containing Li_2O–Al_2O_3–SiO_2 glass-ceramic is dependent upon the ultraviolet radiation dose. Figure 24 shows the variation of modulus of rupture as a function of UV dose for material subsequently heat-treated for 1 hr at 520°C, 1 hr at 620°C and 1 hr at 900°C. The peak strength of about 230 MNm^{-2} was associated with a minimum mean grain size of about 1 μm and, as shown in Fig. 25, there was a clear cut relationship between mean grain size and mechanical strength.

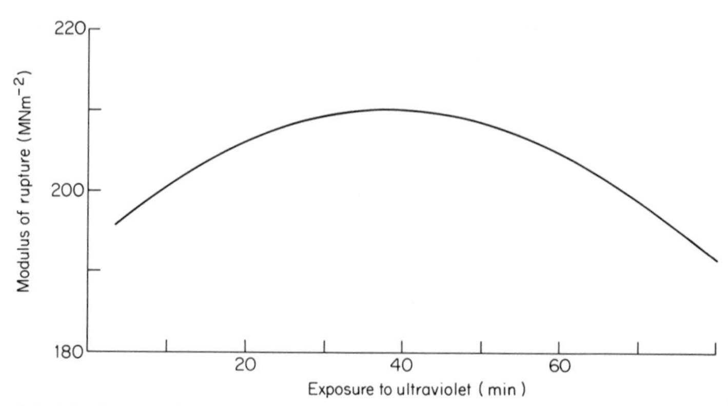

FIG. 24. Modulus of rupture as a function of UV dose for a gold-nucleated Li_2O–Al_2O_3–SiO_2 glass.

The explanation proposed for the existence of an optimum UV irradiation dose was as follows. The number of gold atoms produced in the glass of UV irradiation depends upon the UV dose (ie upon the duration of irradiation). On heat-treatment at 520°C, further reduction of Au^{3+} ions in the glass is nucleated by pre-existing gold atoms and growth of gold particles takes place. For these to be effective subsequently in heterogeneous nucleation, they must attain a minimum critical radius. If, however, the UV dose is excessive, insufficient Au^{3+} ions remain in the glass to permit subsequent growth of all the gold particles to the critical minimum radius. There will therefore exist an

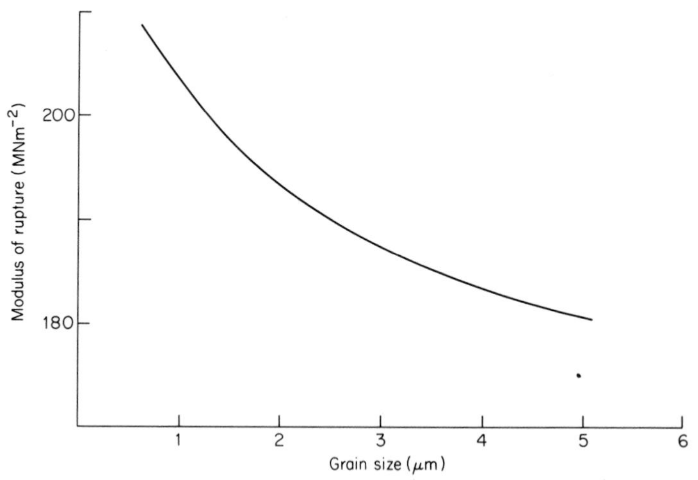

FIG. 25. Modulus of rupture as a function of mean grain size for a gold-nucleated Li_2O–Al_2O_3–SiO_2 glass.

optimum UV dose (radiation duration) which will ensure the maximum concentration of gold particles attaining the critical radius.

2. *Nucleation by Separation of Oxides*

a. Two-phase separation in oxide systems. It was seen in chapter 2 that if the free energy versus composition curve for a two component system exhibits two minima, then the system will tend to separate into two phases for compositions lying between the minima. Though the thermodynamics of two-phase separation are fairly well understood the underlying structural reasons for this behaviour are less clear. Since it appears that many oxide nucleating agents take part in, or induce, phase separation, it is of value to consider the structural effects that are likely to be involved. For reasons discussed earlier, it appears that to be effective, the two-phase separation should take the form of a dispersion of colloidal particles of one phase dispersed in a matrix of the second phase. A number of oxides induce this kind of phase separation in glasses either during cooling of the molten glass or during reheating of the initially chilled glass.

In considering the structural reasons for two-phase separation, it is convenient to consider the effects of network-forming cations separately from those of modifying ions although some ions having an intermediate character may act in both of these roles. The discussion will mainly be confined to examination of effects in silicate glasses although similar considerations may be expected to apply to other glass systems.

The network of a silicate glass is built up of SiO_4 tetrahedra and this can incorporate "foreign" structural units (either tetrahedra or triangles) to develop a composite network structure. At low concentrations of the intrusive units, this structure may be stable so that a single-phase glass is produced. At higher concentrations, the disruptive effect of the foreign structural units may lead to separation into two phases one of which is rich in silica while the other is rich in the added network-forming oxide. If the concentration is raised still further so that the added oxide forms the major component, a single phase system may once more be formed in which the SiO_4 tetrahedra can be regarded as being dissolved in the network of the second oxide. Thus there will exist a range of compositions where two-phase separation is probable provided that some form of incompatibility between the two network formers exists.

There are two basic reasons for incompatibility between different types of network structure. The first of these is based on geometrical considerations. Though the silicate network is a random structure there is evidence that the Si–O–Si bond angle and the Si–O separation do not deviate greatly from certain mean values. The introduction of "foreign" network forming units differing significantly in size from the SiO_4 unit will result in distortion of the Si–O bonding and if the concentration of the added oxide exceeds some critical limit, separation into two phases to lower the total free energy of the system will ensue. Another type of phase separation, also dependent on geometrical effects, can occur when ions are introduced for which the equilibrium coordination number changes with temperature. For a number of cations the coordination number decreases with increasing temperature. In such cases, the equilibrium coordination in the melt could be tetrahedral and such units would be compatible with the silicate network but on cooling the equilibrium coordination becomes octahedral. In such a case, it is possible that phase separation will occur during cooling of the melt or during reheating of the quenched glass.

The second basic reason for incompatibility between different types of network-forming structural groups is that differences between the charge of the principal network-forming ion (silicon) and that of other ions which assume a network-forming position can lead to instability. For example, a tetrahedral unit in which the central cation has a higher charge than the silicon ion may be difficult to accommodate in the silicate network. A possible two-dimensional representation of such a network is given in Fig. 26a. If this arrangement actually occurred, however, it would be highly unstable because electroneutrality cannot be ensured for the regions of the network around the X^{5+} ion. In such a structure each of the XO_4 tetrahedra will bear an excess unit positive charge and it is unlikely that this could be neutralised by interstitial ions since these would have to be negatively charged.

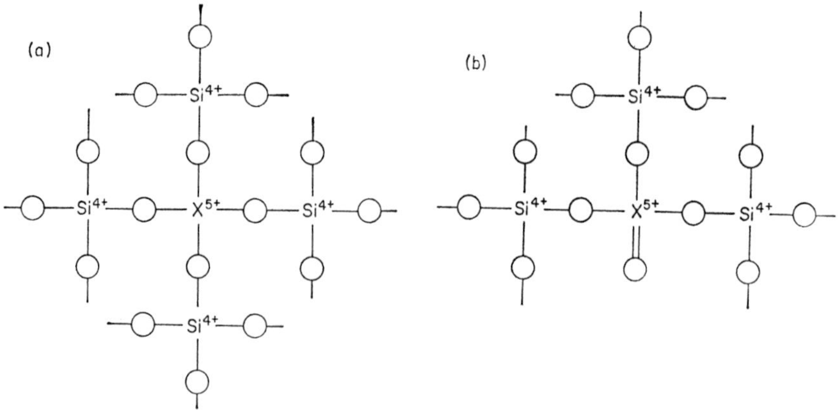

FIG. 26. Possible arrangements of silicate networks containing pentavalent network forming cations. (Two-dimensional representations of the SiO_4 and "XO_4" groups are given for simplicity; the actual network structure is three-dimensional, the SiO_4 and "XO_4" groups having tetrahedral configurations.)

Electroneutrality could be ensured if one of the oxygen ions around the pentavalent ion were doubly bonded to the central cation as shown in Fig. 26b. Such an arrangement might not be particularly stable however since the tetrahedral units containing the pentavalent ion would be asymmetrical and in the regions surrounding these units there would exist marked disturbance of the bonding in the silicate network. Under these circumstances, separation into two phases would be likely to occur.

The effects of interstitial or modifying cations are also important since these may induce two-phase separation. The structure of a silicate glass is a rather open one containing interstices between the linked SiO_4 tetrahedra which can accommodate modifier cations such as alkali or alkaline earth ions. Each type of cation in the glass structure will attempt to achieve the optimum coordination with oxygen. Generally the silicon ions will exert the dominant effect on the overall arrangement of oxygen ions but, depending on their field strengths, the modifier cations will affect the arrangement in attempting to achieve their equilibrium coordination state. Thus a competition process arises in the glass melt and modifier cations of relatively high field strengths may exert a sufficiently large disruption of the silicon oxygen network to result in separation into two phases in order to lower the free energy of the system. Support for this view is given by Tashiro (1966) who, in discussing the work of Dietzel (1949) and Levin (1967) pointed out that the consulate temperatures of binary R_2O–SiO_2 and RO–SiO_2 systems fall linearly with increase in the ionic field strength difference between the modifier cation and silicon. Thus the tendency towards phase separation increases as the field strength of the modifier cation more closely approaches that of silicon.

One of the two phases will be rich in silica and the concentration of modifier cations will be small, the silicon ions largely determining the oxygen ion arrangement. In the second phase there will be a high proportion of interstitial ions and these will exert the predominating influence upon the overall structure. The concentration of non-bridging oxygen ions will be high in this phase, weakening the network structure so that in some cases a coherent silicate network may no longer exist. Thus the modifier rich phase may tend to crystallise more easily than the silica-rich phase.

The extent of immiscibility in binary silicate systems is clearly indicated by the shapes of the liquidus curves. For systems in which two-liquid melts are developed the liquidus temperature remains invariant over a range of compositions corresponding with the immiscibility range, (Levin and Block, 1957). Some systems while not showing a horizontal liquidus line, exhibit S-shaped liquidus curves and this indicates that they will in all probability develop metastable (sub-liquidus) immiscibility. The liquidus curves for the BaO–SiO_2 and Li_2O–SiO_2 systems are distinctly S-shaped while that for the Na_2O–SiO_2 system exhibits this feature to a lesser extent. Glasses of these types are in fact known to exhibit phase separation and the occurrence of this in Li_2O–SiO_2 glasses is of especial interest since these compositions are the progenitors of several important glass-ceramic systems.

b. Practical examples of oxide nucleation catalysts. An important example of an oxide nucleating agent is titanium dioxide the use of which was described by Stookey (1960) who showed that it was effective in a variety of glass compositions in amounts of 2 to 20 weight per cent. At the higher concentrations in this range, titanium dioxide is clearly a major constituent of the glass and it is questionable whether it would be correct to describe it simply as a nucleating agent in such glasses.

Titanium dioxide is soluble in a wide range of molten glasses but during cooling or subsequent reheating large numbers of submicroscopic particles are precipitated and these apparently assist the development of major crystal phases from the glass. The production of the desirable fine-grained microstructure generally involves heating the glass to a temperature about 50°C above its annealing point to accomplish nucleation. It has been found that the glass should be cooled 100°C to 300°C below the nucleation temperature prior to holding at this temperature for a period of 0·5 to 2 hours, in order to achieve optimum nucleation.

The simple idea that crystals of rutile are first precipitated and that these act as heterogeneous nuclei for the crystallisation of other phases is unsatisfactory because rutile does not always appear as a crystal phase and even when it is formed, it is often only detected at late stages in the heat-treatment ie after the precipitation of other crystal phases. There is evidence, however, that the first stage in the process is the separation of a titania-rich liquid (or glass) phase taking the form of a dispersion of very small droplets. This phase separation may be initiated during cooling of the melt, though generally crystal nuclei are not formed at this stage.

Support for the idea that two-phase liquid separation plays an important role in the controlled crystallisation of titania containing glasses was given by the work of Maurer (1962). He used light scattering experiments to investigate microstructural changes in titania-nucleated MgO–Al_2O_3–SiO_2 compositions. The results suggested that two-phase separation, taking the form of an emulsion, had occurred during the early stages of cooling of the molten glass and subsequent heat-treatment had little effect upon the extent of emulsion formation. Crystallisation of the droplet phase was initiated in the temperature range 725°C to 770°C as evidenced by increasing anisotropy of this phase and prolonged heat-treatment at 770°C resulted in a fairly clear X-ray diffraction pattern identified as that of magnesium titanate. Variation of particle sizes from 2·5 nm up to 17·5 nm after prolonged heat-treatment at 770°C was also revealed by the light scattering studies.

It is significant that Maurer's work showed that titanium dioxide did not separate out as the pure compound but as magnesium titaniate, $MgO.TiO_2$. A similar effect occurs in other glasses; for example TiO_2 does not act as a satisfactory nucleating agent in Na_2O–Al_2O_3–SiO_2 glasses unless certain divalent metal oxides are also present. These oxides are those which in combination with TiO_2 form titanates of the general formula $RO.TiO_2$ which are characterised by a close-packed hexagonal structure of the ilmenite type. Suitable oxides include those of iron, cobalt, cadmium, zinc, nickel, manganese and magnesium. It has been observed, also, that TiO_2 does not give satisfactory nucleation in simple Li_2O–SiO_2 glasses, whereas other nucleating agents are effective. The inclusion of MgO and/or Al_2O_3 to give more complex glasses, however, enables satisfactory nucleation with TiO_2 to be achieved.

On the basis of this evidence it appears that in many glass-ceramics, the role of TiO_2 is to form compounds with other oxides and that the compounds so-formed are able to nucleate rather easily by a process involving prior glass-in-glass phase separation.

The process of phase separation itself is related to the structural role of the titanium ion. Titanium dioxide is regarded as having an intermediate effect in glass structures implying that some of the Ti^{4+} ions occupy tetrahedral network-forming sites. In view of its ionic radius (0·68 Å) the titanium ion would normally be expected to assume six-fold co-ordination with oxygen and in most of its crystals it is found in this state. W. A. Weyl (1951) has suggested that at high temperatures the Ti^{4+} will assume four-fold co-ordination with oxygen and in this case the titanium-oxygen groups will be structurally compatible with silicate networks. During cooling, the titanium ion will tend to revert to its low temperature equilibrium co-ordination number (six) and will thus be displaced from the silicate network. In combination with a divalent metal oxide, titanium dioxide is therefore likely to form a separate phase. For some glass compositions the high temperature state of the titanium ions may be "frozen in" but this state would be unstable at low temperatures and reheating of the glass would result in phase separation.

Barry *et al.* (1969, 1970) advanced the idea that titanium dioxide may act in the nature of a surface active agent along similar lines to those first proposed by Hillig (1962). On the basis of experimental investigations of the crystallisation of TiO_2-nucleated glasses of the Li_2O–Al_2O_3–SiO_2 type these authors proposed that because of the limited ability of Ti^{4+} ions to participate in tetrahedral networks, these ions tend to attract non-bridging oxygen ions. Thus they argued that a "domain" structure may arise in the glass in which regions exist in which all of the oxygen ions are bridging: these regions are separated by non-bridging oxygens associated with Ti^{4+} ions. Lithium ions are also envisaged as concentrating at the domain boundaries in association with the non-bridging oxygen ions. For one glass a domain size of 24·3 Å was computed. It was suggested that the effective negative charge on AlO_4 groups would tend to restrain such groups from occupying domain surface sites and therefore all of the Al^{3+} would reside within the domain interiors. The possibility was also proposed that Si^{4+} ions might migrate from the interior of the domains. Depending on the parent glass composition, therefore the compositional differentiation implied by the domain formation could result in enhanced nucleation of various phases.

It should be pointed out that these ideas are based on a large number of assumptions and that direct evidence for domain formation of the type envisaged, is not available.

Titanium dioxide is an example of an oxide where destabilisation of the single phase glass arises because of co-ordination effects. Another intermediate oxide that is likely to exhibit similar behaviour is zirconium dioxide. The use of this oxide as a nucleating agent has been described by Sawai (1961) but it is less useful than TiO_2 first because its solubility in silicate melts is limited to 3 or 4 per cent whereas 20 per cent or more of TiO_2 can be dissolved. A further advantage of the latter is that it markedly lowers the viscosity of the glass melt. Tashiro (1966) found, however, that the solubility of ZrO_2 was significantly increased if P_2O_5 was also included in the glass composition. In this way, improved nucleation using ZrO_2 was achieved but, although this was not realised at the time, the nucleating effect of phosphorus pentoxide itself probably contributed to the improved result.

The use of phosphorus pentoxide as a nucleating agent was discovered by McMillan and Partridge (1963a). This oxide provides an example of a network-former which exhibits the characteristics of a nucleating agent. The phosphorus ion, P^{5+}, assumes tetrahedral co-ordination and therefore provides an example of phase separation due to a charge difference between the principal network-forming ions, Si^{4+}, and the "foreign" network-forming ions, P^{5+}. As we have seen (Fig. 26), if the phosphorus-oxygen bonds were all single bonds of the P–O type, electroneutrality could not be ensured so that one phosphorus–oxygen bond per PO_4 tetrahedron would have to be a double bond. The presence of this type of double-bonded oxygen ion within the silicate network creates conditions favouring separation of phosphate grouping from the silicate network. It is unlikely that these groups would separate as phosphorus pentoxide but that this oxide would separate out in combination with an alkali or alkaline earth oxide. The phase which separates out may be crystalline but it is quite likely to be glassy. Electron microscopy of glasses containing phosphate-induced two-phase separation does not reveal any evidence of crystalline regularity within the phase which has separated and neither does X-ray diffraction analysis. In certain types of phosphate opacified glass, however, definite crystals of calcium phosphate have been identified but these were soda-lime-silica glasses, quite different in composition from the glasses in which phosphate nucleation has been utilised. Quite small concentrations of P_2O_5 are effective in inducing the desired two-phase separation necessary for nucleation catalysis and amounts ranging from 0·5 to 6 weight per cent have been used successfully although a concentration of 3 weight per cent appears to represent the optimum proportion. McMillan and Partridge (1963a) have described the use of metallic phosphates to catalyse the crystallisation of a wide range of glass compositions derived from the Li_2O–Al_2O_3–SiO_2, Li_2O–MgO–SiO_2 and MgO–Al_2O_3–SiO_2 systems. The same nucleation method has also been successfully applied to Li_2O–ZnO–SiO_2 glasses and to Li_2O–PbO–SiO_2 glasses.

James and McMillan (1968, 1970) have investigated the relationship between prior phase separation and crystal nucleation by undertaking high resolution transmission electron microscopy of glasses of the Li_2O–SiO_2 type to which small additions of P_2O_5 introduced as lithium phosphate were made. It was found that a glass of the molecular percentage composition Li_2O 30, SiO_2 70 underwent extensive phase separation into silica-rich droplets in a lithia-rich matrix after heat treatment at temperatures in the region of 550°C (Fig. 27). Substitution of 1 mol per cent P_2O_5 for the equivalent amount of SiO_2 led to a change in the morphology of phase separation (Fig. 28), to an increase in the numbers of particles per unit volume by a factor of about 1·5

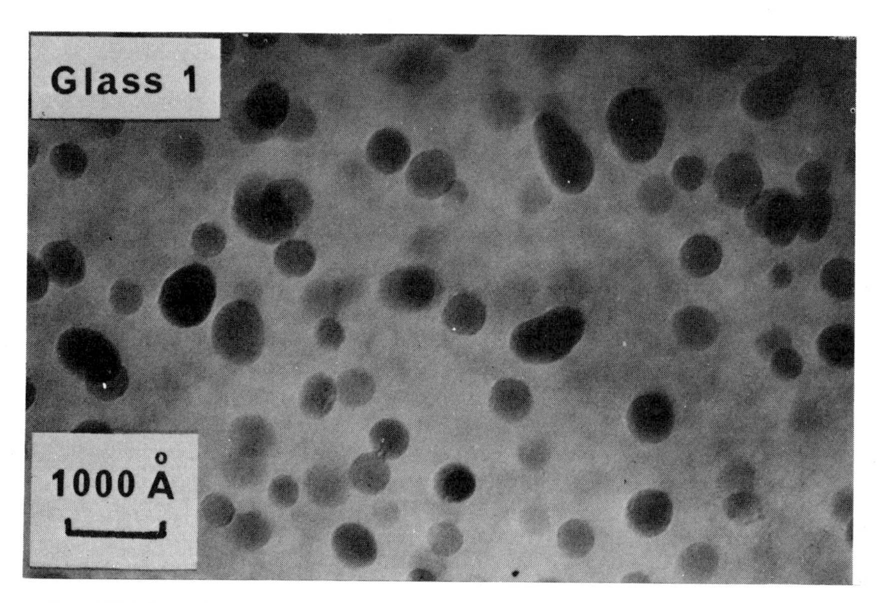

FIG. 27. The microstructure of a glass of the molecular percentage composition SiO_2 70; Li_2O 30 after heat-treatment at 550°C for 1 hr.

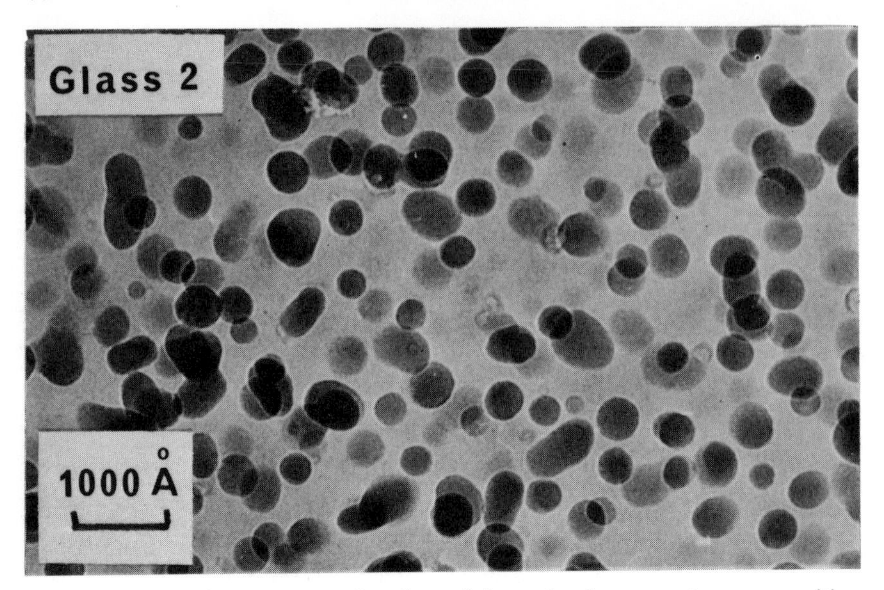

FIG. 28. The microstructure of a glass of the molecular percentage composition SiO_2 69; Li_2O 30; P_2O_5 1 after heat-treatment at 550°C for 1 hr.

and to an increase in the volume fraction of the particulate phase by a factor of about 2·5. Various possible models to explain the significant influence of a small concentration of P_2O_5 were examined. The model which gave the best fit between calculated and observed volume fraction of dispersed phase was one in which the matrix phase had a composition slightly richer in Li_2O than the disilicate composition $Li_2O.2SiO_2$. It was also deduced that the P_2O_5 was present in the matrix phase and that it was associated with Li_2O, probably in accordance with the lithium orthophosphate composition $Li_2O.3P_2O_5$. Measurements of dielectric loss for the phase-separated glasses gave results that could be explained in terms of this model. The exact form of the lithium orthophosphate in the glass was not resolved. No X-ray diffraction pattern for this compound was obtained so that if crystals were present they were of an insufficient size to yield a coherent diffraction pattern (ie they were less than ~ 10 nm in size).

Confirmation that P_2O_5 enhances phase separation in Li_2O–SiO_2 glasses was given by the work of Tomozowa (1971) who showed that the substitution of 1 mol per cent P_2O_5 for SiO_2 in compositions from the Li_2O–SiO_2 system raised the immiscibility temperatures and displaced the immiscibility boundary towards higher Li_2O contents. A similar effect was apparent for the Na_2O–SiO_2 system. Matusita *et al.* (1975) undertook a comprehensive study of phase separation and crystallisation of glasses based on the $Li_2O.3SiO_2$

composition. Various oxides were added to the base glass in the proportion of 3 molecular per cent. It was noted that P_2O_5 increased the immiscibility temperature of the glass by about 200°C whereas TiO_2 addition resulted in only a relatively small increase. The outstanding effect of P_2O_5 was attributed by the authors to the high field strength of the phosphorus ion.

The tendency for two-phase separation to occur in a melt or glass containing phosphorus pentoxide is reduced if the glass contains a high proportion of aluminium oxide. This effect is explainable in terms of the roles of the aluminium and phosphorus ions in glass structure. The aluminium ion occupies the centre of a tetrahedral AlO_4 group which takes part in the glass network and this group bears an excess unit negative charge. In many glasses this charge is neutralised by the presence of alkali or alkaline earth ions but when phosphorus is present in the glass the AlO_4 group can achieve electro-neutrality by becoming linked to a PO_4 group as shown in Fig. 29. In this arrangement, the excess charge of +1 associated with the PO_4 group serves to neutralise the excess charge of −1 associated with the AlO_4 group and there is no necessity for one of the oxygen ions linked to the phosphorus to be doubly bonded to it in order to ensure electroneutrality. In addition to this, the radii of the aluminium and phosphorus ions (0·50 Å and 0·34 Å respectively) are such that when these ions occupy the centres of adjacent tetrahedral groups, the overall atomic spacing for the two groups will be quite close to that for an adjacent pair of SiO_4 groups in which the central ion, Si^{4+}, has radius of 0·41 Å. Thus both from the point of view of ensuring electroneutrality and from

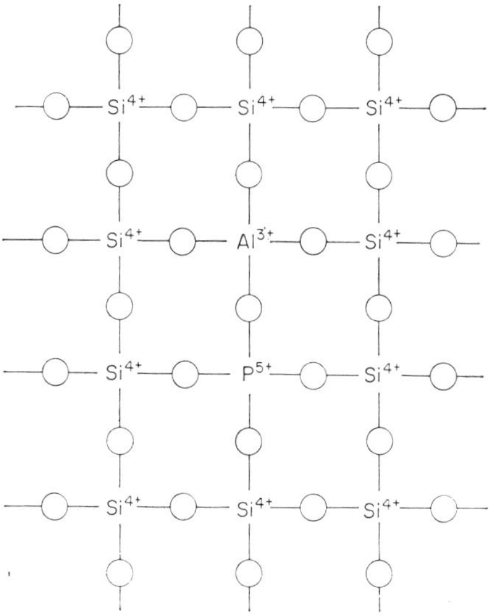

Si4+ Si4+ Si4+
Si4+ Al3+ Si4+
Si4+ P5+ Si4+
Si4+ Si4+ Si4+

FIG. 29. Schematic arrangement of AlO_4 and PO_4 groups in a silicate network.

considerations of the size of tetrahedral groups participating in the network, linking together of AlO_4 and PO_4 groups would tend to stabilise the phosphate tetrahedra within the silicate network. Additional evidence on this point is given by the observation that crystalline aluminium orthophosphate $AlPO_4$ bears a strong structural resemblance to a crystalline form of silica (cristobalite) and the stability of AlO_4–PO_4 groups within the silicate network would therefore be expected.

Vanadium pentoxide (V_2O_5) may, in small concentrations, play a similar structural role in glass to P_2O_5. The vanadium ion, with a radius of 0·59 Å is larger than ions which normally form stable tetrahedral groups with oxygen but V_2O_5 is a glass-forming oxide and various vanadate glasses have been prepared. The nature of the co-ordination polyhedra in these glasses has not been firmly established. It may be tetrahedral but the co-ordination polyhedra in the crystalline oxide is a greatly distorted octahedron and the structure is best described as composed of sheets of distorted trigonal bipyramids. If, as seems likely, vanadium pentoxide assumes a network-forming role in silicate glasses the additional positive charge of the V^{5+} ion as compared with neighbouring Si^{4+} ions may induce an instability leading to microphase separation. It has been shown that V_2O_5 can act as a nucleating agent, though considerably less effectively than P_2O_5, in glasses basically of the Li_2O–SiO_2 type.

A complication that occurs with vanadium but not with phosphorus is that the former can exist in several valence states in glasses. In fact it would be unlikely that the whole of the vanadium content of a glass would be present in the pentavalent form. During melting, loss of oxygen can occur with the result that V^{4+} and V^{3+} ions may be formed. Thus the control of the nucleation process in glasses containing vanadium would be rendered difficult.

Williamson *et al.* (1969) studied the effects of both V_2O_3 and V_2O_5 additions on the crystallisation of CaO–MgO–Al_2O_3–SiO_2 glasses. They concluded that in these glasses neither oxide promoted satisfactory internal nucleation. Effects upon the growth rates of surface-nucleated anorthite and wollastonite crystals were observed however. Both oxides tended to enhance the growth rates but the effect was greater for V_2O_5 additions.

As we have seen from the earlier discussion (pp. 73–4), certain interstitial cations of high field strength will tend to cause a disturbance of the silicate network due to their strict requirements for achieving the equilibrium co-ordination number. Those ions will therefore induce two-phase separation or crystal nucleation in the melt or glass.

Chromium is an example of a metal which can exist in glass in the form of an interstitial cation of high field strength. For glasses melted under oxidising conditions chromium can exist in glass in two valency states, namely as the trivalent chromic ion Cr^{3+} and as the hexavalent Cr^{6+} ion. The relative proportions of the two types of ion are determined by the melting conditions and the glass composition. For example, increased melting temperatures or prolonged melting appear to favour the formation of trivalent chromium. The presence of alkali metal oxides in the glass favours the formation of the hexavalent state and potassium oxide exerts a stronger influence in this respect than lithium oxide, sodium oxide having an intermediate effect. Lead oxide also appears to favour the development of hexavalent chromium. Divalent chromous Cr^{2+} ions can be present in glasses melted under strongly reducing conditions.

Chromium in the hexavalent state Cr^{6+} represents an ion with a high field strength ($Z/r^2 = 17{\cdot}2$) and since it is likely that it would occupy an interstitial position, this ion would exert a marked "ordering" effect upon the oxygen ions which surround it. Under these circumstances, a chromium-rich phase would separate out from the glass. Evidence that chromium can behave in this manner is found for certain types of lead silicate glaze in which lead chromates crystallise out, imparting yellow or red colours to the glaze. In other types of glasses, chromic oxide can be induced to crystallise out in the form of relatively large thin plates to give the chromium aventurine, a glass at one time used for decorative purposes but now rarely made.

Sawai (1961) has reported work carried out in Japan in which 1 to 3 per cent of chromic oxide was successfully used to catalyse the crystallisation of an aluminosilicate glass containing lithia, potash and calcium oxide. This provides a practical demonstration that Cr_2O_3 can be employed to induce two-phase separation in glasses followed by crystallisation of other phases.

On the other hand, Thakur (1971) reported that Cr_2O_3 did not act as an effective nucleating agent for the bulk crystallisation of $Li_2O.2SiO_2$ glass. His results appeared to show that this oxide resulted in enhanced surface nucleation. The same author, however, stated that a complex glass of the K_2O–MgO–CaO–Al_2O_3–SiO_2 type could be successfully converted to a glass-ceramic using a complex nucleating agent comprising Fe_2O_3, Cr_2O_3 and CaF_2.

Williamson (1970) found that chromic oxide was not effective as a nucleating agent in a CaO–MgO–Al_2O_3–SiO_2 glass but one difficulty was the low solubility (0·6 per cent) of the oxide in the glass studied.

Shelestak *et al.* (1978) investigating the production of a glass-ceramic from a glass made by melting oil shale found that the parent glass exhibited only surface crystallisation. By the addition of about 1 per cent chromic oxide, however, they were able to produce a material that was approximately 90 per cent crystalline with an average grain size less than 10 μm. In this case the nuclei formed were found to be a mixed spinel type of crystal of the general formula (Mg, Fe) (Al, Fe, Cr)$_2$O$_4$.

Like chromium, molybdenum also forms a hexavalent ion, and it is likely that molybdenum can exist in this state in glass, although depending upon the conditions of melting and upon the composition of the glass other ions such as the Mo^{4+} ions may also be present. There is little direct evidence concerning the possible states of valency of molybdenum in glass, however. In its compounds molybdenum can exhibit six different oxidation states 0, +2, +3, +4, +5 and +6, and it might be expected that a series of oxides would exist corresponding with valencies from 2 to 6. Only two oxides are completely confirmed as true chemical entities; these are MoO_2 and MoO_3. It is reasonable to suppose, therefore, that in glass the likely states of the molybdenum will be as the tetravalent Mo^{4+} and the hexavalent Mo^{6+} ions. The hexavalent ion, with a radius of 0·62 Å, has a high field strength ($Z/r^2 = 15{\cdot}6$) and a molybdenum-rich phase might separate from the glass rather easily since the interstitial molybdenum ions would exert a strong ordering effect upon the surrounding oxygen ions. Tungsten, which also forms a hexavalent ion of relatively high field strength, would behave in a similar fashion. The solubilities of molybdenum and tungsten oxides in glasses are not high but amounts up to 3 or 4 per cent can be incorporated in certain types of glasses. Both oxides cause opacification of glasses due to the separation of a crystalline phase, and this effect has been made use of in the manufacture of enamels and vitreous colours based on glasses of high lead oxide content. It would be expected that molybdenum and tungsten oxides could be used as nucleation catalysts since they exhibit some of the necessary characteristics for this purpose. The oxides are soluble in glass melts at high temperatures and for some types of glass composition they can remain in solution during cooling of the glass melt. Reheating of glasses containing these oxides causes fine crystalline precipitates to appear at temperatures in the region of the softening temperature of the glass. These crystals could then be used to nucleate the crystallisation of other supersaturated phases from the glass.

Molybdenum or tungsten oxides have been used as crystallisation catalysts for the production of glass-ceramics by McMillan and Partridge (1961). It was shown that the controlled crystallisation of a wide range of lithia-containing glasses, including compositions of high alumina content, could be achieved by the inclusion of 0·5 to 4 per cent of MoO_3 or WO_3 in the glasses. The heat-treatment of the glasses included a nucleation stage at a temperature near to the dilatometric softening point ($\log_{10}$ viscosity $= 10^{11}$ to 10^{12} poises) of the glass followed by further heat-treatment to crystallise major amounts of various compounds.

The oxides of iron have proved to act as useful nucleating agents in some glass-ceramic systems. Williamson *et al.* (1968) showed that the kinetics of growth of anorthite and wollastonite from a $CaO–MgO–Al_2O_3–SiO_2$ glass were affected by the inclusion of iron oxide in the materials. Crystal growth rates depended on the concentration of iron oxide but the ferrous/ferric ratio was also important. In fact the crystal growth rate was found to increase linearly as the concentration of Fe^{2+} ions was raised. Although these ions were found to exert a major influence upon crystal growth processes, they did not appear to affect internal nucleation. A tendency towards internal nucleation was only found when the iron oxide concentration present mainly in the ferric form, exceeded 5 weight per cent.

In a later study, Rogers and Williamson (1969) investigated nucleation in the same type of glass to which ferric oxide (Fe_2O_3) or ferrous oxalate ($Fe(COO)_2\ 2H_2O$) were added. It was found that glasses containing upwards of 7 weight per cent of Fe_2O_3 could be induced to nucleate internally and to grow a large number of small dendritic crystals. The main crystal phases were either diopside or melilite. The latter comprises a series of solid solutions between gehlenite ($2CaO.Al_2O_3.SiO_2$) and akermanite ($2CaO.MgO.SiO_2$). Diopside predominated in glass that had been crystallised for short periods while melilite grew in glass heat-treated for long periods at lower temperatures ($\sim$950°C). The crystallised glasses often contained traces of a spinel phase for which the unit cell was slightly larger than the normal spinel, $MgAl_2O_4$. It was suggested that this indicated partial replacement of Al^{2+} ions by Fe^{3+}. Analysis of the lattice parameters indicated an Al_2O_3:Fe_2O_3 ratio of 12:1 which was about four times higher than the ratio in the parent glass. It was hypothesised that the ferric ions in the glass largely occupy tetrahedral sites and that in so doing they tend to displace Al^{3+} ions. The latter are thereby induced to take up octahedral co-ordination and are thus more favourably disposed for the production of spinel nuclei. These nuclei provide growth sites for diopside and melilite crystals. Thus the action of ferric iron in prompting crystal nucleation in the glass was seen as a complex one.

A further example of the action of ferric ion in promoting crystal nucleation in glasses was given by the work of Beall and Rittler (1976) who studied the formation of glass-ceramics from glasses of the basalt type made by melting the crushed rock. It was established that the most important factor controlling nucleation and hence the microstructure of the basalt glass-ceramics was the state of oxidation of the iron. The starting material contained 4·4 wt per cent Fe_2O_3 and 8·4 wt per cent FeO and thus the iron was predominantly in the ferrous state. During melting, the equilibrium tends to shift in the direction of the ferric form and the extent of this reaction is determined by factors such as melting temperature, raw material particle size, batch additions and bubbling of gases. For example, using standard melting conditions of 1400°C for 4 hours in air it was found that the ratio FeO:Fe_2O_3 was 1·9 in glass prepared from granules. Use of finely powdered raw material caused the ratio to fall to 1·6 and addition of 4 per cent NH_4NO_3 resulted in a further reduction of the ratio to 0·5. The latter value approaches the ratio in magnetite (Fe_3O_4).

The strong influence of the state of oxidation of the iron upon nucleation efficiency was shown by comparing the crystallisation behaviour of two basalt glasses to one of which a reducing agent (2 per cent sugar) and to the other an oxidising agent (4 per cent NH_4NO_3) had been added. Heat treatment of the reduced glass gave a microstructure comprising large ($\sim$ 50 μm) spherulites of clinopyroxene in a glass matrix; deformation of the material occurred during heat treatment. The oxidised glass gave a fine-grained pyroxene glass-ceramic and crystal sizes of $\sim$ 100 nm could be achieved.

It was established that a nucleation treatment of 2 hours at temperatures in the range 650–700°C resulted in the precipitation of magnetite nuclei. Further heat treatment at 900°C caused the precipitation of pyroxene crystals onto these nuclei. Thus the need for oxidising the glass is made clear; the ratio of Fe^{3+} to Fe^{2+} ions must be close to that in magnetite to allow the initial development of the nuclei.

It is interesting to note that if TiO_2 is present in the basalt glass, fine-grained glass-ceramic microstructures can be produced even from the chemically reduced glasses. The reason is that TiO_2 combines with FeO to form a spinel Fe_2TiO_4 which is structurally equivalent to magnetite and which therefore will also provide the required form of nucleation.

So far, the oxide nucleating agents discussed have generally comprised a single component included in a crystallisable glass. It has been discovered, however, that in some cases mixtures of two oxides, either of which will act as a nucleating agent when used alone, give improved nucleation. A synergistic effect occurs so that for the same overall concentration of nucleant, the mixture gives a higher nucleation density and hence a finer-grained microstructure than the single nucleants. McMillan and Partridge (1966a) used mixtures of 0·5 to 6 per cent P_2O_5 combined with 0·2 to 0·9 per cent TiO_2, 0·5 to 4·0 per cent WoO_3, 0·4 to 4 per cent MoO_3 or 0·5 to 2·0 per cent V_2O_5. In all cases improved nucleation resulted from the use of mixed nucleating agents. The same investigators (1966b) discovered that mixtures of MoO_3 and WoO_3 in the weight percentage range 0·5 to 4·0 also gave improved nucleation. McMillan and Lawton (1967) reported a similar effect for mixtures of MoO_3 and ZrO_2. Preferably, these were introduced in approximately equal proportions to give a total of 10 weight per cent. Stewart (1971) discovered that mixtures of TiO_2 and ZrO_2 in a total concentration of 2·23 molecular per cent gave glass-ceramics having a finer grain size and higher mechanical strength than similar materials using either of the two oxides alone. The glass-ceramics containing the mixed nucleating agent were also less sensitive to variations of the heat-treatment cycle.

So far, no explanation of the enhanced nucleation resulting from the use of mixed oxides has been advanced. Possibly one of the oxides promotes nucleation while the other favourably modifies the crystal growth characteristics but this topic awaits a thorough experimental study.

The addition of fluorides to glasses of the soda-lime-silica type to opacify them has long been known and compounds which have been used for this purpose include sodium fluoride (NaF), cryolite (Na_3AlF_6) and sodium silico-fluoride (Na_2SiF_6). In small concentrations (less than 1 per cent) the fluorine can be accommodated in the glass structure and clear glasses can be obtained. With higher amounts (greater than 2 to 4 per cent) fluorides come out of solution during cooling of the glass with the result that a fine crystalline precipitate is formed so that the glass is opaque. It is possible to adjust the amount of fluorine present in a glass so that the glass remains clear on cooling, but on reheating to temperatures slightly above its annealing range, precipitation of the fluoride crystals occurs. From this it is clear that it might be possible to make use of fluoride opacification of glasses as a means of catalysing the crystallisation of glasses, the tiny fluoride crystals acting as nucleation centres. Chlorides also can cause opacification of glasses, although usually only very small proportions can be dissolved in silicate glass melts so that it is perhaps unlikely that these compounds will find practical applications as nucleating agents.

It is of interest to consider how the fluorine ion can take part in the structure of a silicate melt or glass. The fluorine ion, with a radius of 1·36 Å, is very close in size to the oxygen ion (radius 1·40 Å) so that a fluorine ion could replace an oxygen ion in the glass network without causing too great a disturbance in the arrangement of other ions. Since the fluorine ion is monovalent, however, the overall replacement within the glass must be of two fluorine ions for each oxygen ion in order to ensure electroneutrality. A general effect of the introduction of fluorine would be to replace strong Si–O–Si linkages by pairs of Si–F linkages with the result that the glass network structure would be weakened. This effect is reflected in decreased viscosity of the melt and by a tendency for the thermal expansion coefficient of the glass to be increased. Complete replacement of oxygen by fluorine cannot be achieved in silicate glasses of course, since the network structure would be completely disrupted.

The ease with which fluorides can be induced to crystallise out of a glass can be attributed to the weakening effect on the glass network. This means that even if the melt can be cooled to give a clear glass, reheating of the glass to a temperature within or just above the annealing range permits atomic rearrangement to occur so that crystalline fluoride nuclei may be formed. The nuclei are formed at a temperature where the growth rate of the fluoride crystals is low so that the fluoride is precipitated as a large number of very fine crystals rather than as a few large crystals.

Has and Stelian (1960) have described the use of metallic fluorides to assist the crystallisation of glasses. The glasses are of the alumino-silicate type and contain alkali metal oxides (Na_2O and K_2O) together with various alkaline earth oxides. The proportion of fluoride used corresponds to 3 to 8 per cent of fluorine in the glass and crystallisation is accomplished by heat-treatment at temperatures within the range 600°C to 1000°C. Fluoride crystals are first precipitated and these serve as nucleation sites for the growth of silicate crystals so that the glass is converted to a polycrystalline glass-ceramic. Metallic fluorides have also been used as nucleating agents in glasses containing major proportions of lithia (McMillan and Partridge, 1959a). Amounts of fluorine ranging from 0·5 to 2·5 weight per cent were employed and the glasses could be converted into microcrystalline glass-ceramics by a suitable heat-treatment process.

A number of metallic sulphides can be dissolved in molten silicate glasses and they may remain in solution when the glass is cooled so that a transparent glass is obtained. If such glasses are reheated to a temperature near to the upper annealing temperature, a colour develops by the process known as "striking". Glasses containing cadmium sulphide provide excellent examples of this effect since they are practically colourless after melting and cooling but develop a rich yellow colour when they are reheated. It seems likely that the development of the colour is due to the production of a colloidal precipitate of cadmium sulphide. If increasing amounts of selenium are added to a glass containing cadmium sulphide, the yellow colour is replaced first by an orange colour and finally by a brilliant red colour known as the selenium ruby. The colouring agents in such glasses appear to be crystals of colloidal dimensions. These crystals can be described as cadmium sulpho-selenides, representing members of a series of mixed crystals between cadmium sulphide (CdS) and cadmium selenide (CdSe).

Since sulphides and sulpho-selenides can be obtained in the form of colloidal precipitates in glasses, this suggests that these compounds might be of value as nucleation catalysts for the production of microcrystalline glass-ceramics at least for certain types of glass composition. It has been demonstrated that silicate glasses containing lithia, magnesia and alumina as major constituents can be converted into microcrystalline glass-ceramics by the inclusion of 0·5 to 2 weight per cent of cadmium sulpho-selenide in the glass compositions (McMillan and Partridge, 1959b). After preparation, the glasses are reheated to a temperature just above the dilatometric softening temperature and are held at this temperature for about 1 hour. This treatment serves to cause "striking" of the cadmium sulpho-selenide and the colloidal particles nucleate the crystallisation of lithium silicates and other crystals. Crystallisation of the glass is carried further by raising the temperature to an upper value which is dependent upon the composition of the glass being processed but is generally in the range of 800°C to 1000°C. By this means, microcrystalline ceramics are obtained which have high mechanical strengths. One disadvantage of cadmium sulpho-selenide as a nucleation catalyst is that the glasses must be melted under carefully controlled reducing conditions to prevent "burning out" of the sulpho-selenide due to oxidation. In addition, the final glass-ceramic is coloured a deep yellow or orange-yellow and this may not always be acceptable.

B. Selection of Glass Types

In selecting compositions for glass-ceramic production there are a number of important factors which have to be taken into account and it is the purpose of this section to deal with these.

1. *Behaviour of Glasses during Melting and Shaping*

The glass compositions must be capable of being melted and shaped by economic methods so that in the formulation of glasses for glass-ceramic production the influence of composition upon these factors must be borne in mind.

a. Melting characteristics. It is important that the melting temperature of the glass shall not be excessively high and, generally speaking, 1600°C would be regarded as an upper limit for practical operations. Excessively high temperatures result in problems concerning the glass furnace refractories and for this reason glass compositions which have sufficient fluidity to permit them to be melted and refined at temperatures not higher than 1400 to 1500°C will be chosen wherever possible. Certain glass constituents lower the viscosity of the glass melt and are therefore valuable in speeding up melting and refining (removal of gas bubbles). The alkali metal oxides have this effect and are to be regarded as useful fluxes in the melting process. Lithium oxide has a greater effect in reducing the viscosity of the melt than sodium oxide which, in turn, has a greater effect than potassium oxide, the comparisons being made for equimolecular proportions of the oxides. The alkaline earth oxides also have useful fluxing effects as do zinc oxide and lead oxide. Boric oxide has a particularly useful effect in decreasing the time needed to achieve satisfactory melting and refining. Some of these fluxes may be used in minor proportions and this would usually be the case for boric oxide but in some cases the oxides concerned may form major constituents of the glass-ceramics. It is useful to note that titanium dioxide which acts as an effective-nucleating agent in a number of glass systems, has a marked fluxing effect and allows economic melting temperatures to be achieved for what would otherwise be quite refractory compositions.

In addition to oxides which have a beneficial effect on melting characteristics, there are those which increase the difficulty of melting by increasing the viscosity of the glass. Alumina is an example of such an oxide but since the presence of this oxide is necessary for certain important types of glass-ceramic, the lower melting and refining rates of glasses containing alumina have to be accepted.

Apart from the ease of melting, the ability to reproduce the glass composition is also important. Variations in the chemical composition can arise from two effects. The first of these is loss of volatile constituents from the glass during melting, and the second factor is change of composition due to solution of the refractory materials in contact with the molten glass. Only in exceptional cases are the losses due to volatilisation troublesome, since at normal melting temperatures silica, alumina and the alkaline earths have low vapour pressures, and while it is true that alkali metal oxides are volatile, the losses are usually small and can be allowed for. Changes of composition due to corrosion of refractories can be serious for some glasses and expensive refractory materials, or even the use of platinum-lined furnaces, may be necessary. Glasses containing high proportions of alkali or alkaline earth oxides are often highly corrosive and melts containing high proportions of lithium oxide lead to rapid attack upon the more usual types of refractory material. The inclusion of fluorides in the melt leads to a marked increase in the rate of solution of the refractory material since fluorine reacts with silica to form the volatile silicon tetrafluoride. Loss of silica from the body of the glass-melt can also occur by the same mechanism.

An important aspect of the chemical stability of glasses during the melting process concerns the possibility of chemical reduction of certain glass constituents due to reaction with the furnace atmosphere. Ideally, the glass composition would not be sensitive to the furnace atmosphere since this would simplify melting techniques. This is not always possible, however, and if it is essential to include reducible oxides such as lead oxide, melting under oxidising conditions is necessary.

It should be pointed out here that the state of oxidation of titanium can be affected by the nature of the melting furnace atmosphere. If reducing conditions prevail, Ti^{3+} ions can be formed, as evidenced by the development of bluish colours in the glass; when melted under oxidising conditions glass compositions containing TiO_2 have a brownish-amber colour. Partial reduction of the titanium content results in impaired nucleation efficiency and hence to coarser-grained glass-ceramics having lower mechanical strengths. Other nucleating agents such as the oxides of iron, molybdenum and tungsten are also sensitive to melting furnace atmosphere and control of the state of oxidation is therefore necessary if controlled nucleation is to be achieved. On the other hand, phosphorus pentoxide is quite insensitive to melting furnace conditions because phosphorus in glass exists in only one valence state. Hence glass-ceramic compositions containing this nucleating agent can be melted under a range of conditions with no necessity for furnace atmosphere control.

b. Working characteristics. The working characteristics of a glass are of great practical importance since they determine which shaping processes can be applied or, indeed, whether any shaping process other than a simple gravity casting technique can be used. Many glass shaping processes depend on the fact that glasses are fluid or plastic over a fairly wide temperature range. In glass manufacture the molten glass is removed from the furnace at a viscosity of about 10^3 poises and the temperature corresponding with this viscosity defines the upper end of the working range. The lower end of the working range is the temperature at which the glass becomes too stiff to be shaped and at this temperature the glass has a viscosity of about 10^8 poises. For many glass-shaping operations a long working range is desirable to give adequate time for the flow or deformation of the hot glass as it cools after removal from the furnace. There are cases, however, where a long working range is not essential or desirable. For example, in the production of sheet glass by a drawing operation, a quick-setting glass is preferred.

The working ranges of many of the glasses suitable for glass-ceramic production tend to be shorter than those for normal glasses. This is a result of the unusual compositions of glass-ceramics and for certain materials the type of shaping operation has to be restricted to those which can accommodate quick-setting glasses; these processes include gravity and centrifugal casting. The presence of alkali metal oxides in the glass is desirable to enhance the working range and glasses which are alkali free, especially those which contain high proportions of calcium or magnesium oxides, tend to have undesirably short working ranges. Glasses with high aluminium oxide contents effectively have short working ranges; this is because such glasses have high viscosities with the result that the temperatures at which shaping is carried out are high. Under these circumstances the glass cools quickly so that it sets rapidly. The inclusion of lead oxide and, to a lesser degree, zinc oxide improves the working characteristics of glasses. In many cases, the rather restricted working characteristics of the glasses are accepted because the chemical compositions have been formulated to permit the development of desirable characteristics in the final glass-ceramic.

Another important aspect of the working characteristics concerns the possibility of devitrification during the cooling of the glass in the shaping operation. Although the primary object in the production of a glass-ceramic is to devitrify the glass, this process must be carried out in a controlled manner. Crystal growth due to homogeneous nucleation during cooling of the glass leads to the production of large crystals and not to the fine microcrystalline structure which can be achieved by catalysed crystallisation. Uncontrolled crystallisation is not therefore likely to enable high-strength glass-ceramics of controlled properties to be produced. In addition, crystal growth during the shaping operation will adversely affect the working properties since it will be accompanied by sharp changes of the viscosity. Also the large crystals produced may lead to the generation of prohibitively high stresses which could result in fracture of the glass articles. In selecting glass compositions these factors have to be taken into account, although crystallisation during cooling or working of the glass could be avoided in certain cases by the adoption of special glass-shaping techniques which would permit the glass to be cooled rapidly through the critical temperature range. Glasses containing high proportions of alkali metal oxides tend to devitrify rather readily during working of the glass, and both lithium and magnesium oxides are especially likely to accentuate the tendency of a glass to devitrify. Fortunately, quite small additions of certain oxides tend to suppress devitrification during cooling of the glass and these can be used to improve the working characteristics in this respect. In many cases the effect of the additional oxides is to lower the liquidus temperature of the glass so that this no longer occurs in the temperature range where the glass is shaped. Aluminium oxide has a marked effect and even a few per cent will suppress devitrification during shaping of the glass. Zinc oxide has a similar but less marked effect in certain glasses and small additions of boric oxide can also be useful. In lithia-containing glasses, small additions of sodium or potassium oxide often have a beneficial effect.

A particularly useful and interesting observation is that small additions of phosphorus pentoxide suppress crystallisation of glass-ceramic compositions derived from the lithia-silica system at temperatures within the glass-working range. This results from the marked effect of P_2O_5 in reducing crystal growth rates in this temperature range as discussed in chapter 2. Vanadium pentoxide would be expected to have a similar but less marked effect.

2. *Chemical Stability of Glasses*

The chemical stability of a glass-ceramic under various conditions is obviously of great importance and the chemical composition of the parent glass will strongly influence this characteristic. Although the factors which govern the chemical stability of glasses are fairly well known, there is little information concerning this aspect of glass-ceramics. There may be present in a glass-ceramic several different crystalline compounds together with a residual glass-phase and the relative resistances of these phases to attack by water or other reagents will determine the chemical stability. It is very probable that the chemical stability of the residual glass phase will play an important part in determining the chemical stability of the glass-ceramic as a whole. It is also probably true that a glass which exhibits poor chemical stability is unlikely to give rise to a glass-ceramic of high stability. To this extent, therefore, the factors which govern the stability of glass-ceramics can be equated with those which determine the chemical stability of glasses. Applying this principle, glasses containing high proportions of sodium or potassium oxides will be avoided since such glasses do not exhibit high chemical stability. Furthermore, it is likely that a major part of the Na_2O and K_2O will appear in the residual glass-phase of the glass-ceramic and will therefore impair its chemical durability. When comparisons are made on an equimolecular basis, lithium oxide has a less serious effect upon the chemical durability than the other alkali oxides, so that its presence as a major constituent of the glass-phase of the glass-ceramic is acceptable. Nevertheless, it is desirable to avoid compositions containing very high proportions of lithium oxide. As pointed out earlier, the inclusion of alkali metal oxides other than lithium oxide may be desirable in some glass-ceramic compositions. In such cases it may be possible to design the composition to make use of the "mixed alkali effect" whereby a mixture of two alkali metal oxides confers greater chemical stability than the same total concentration of either oxide alone.

The alkaline earth oxides such as MgO and CaO increase the chemical stability of glasses and are therefore valuable constituents in a glass-ceramic. Similarly, both alumina and zinc oxide have beneficial effects upon the chemical stability. The inclusion of small proportions of boric oxide can be beneficial in some compositions since this oxide tends to be concentrated in the residual glass-phase and would improve its chemical stability. Usually, the proportions of boric oxide would be limited to 2 to 3 per cent since this oxide will lower the refractoriness of the glass phase so that deformation of the glass-ceramic could occur during heat-treatment.

For certain applications, it is important that the glass-ceramic shall be unaffected by contact with reducing gases at high temperatures. In such cases the composition of the glass-ceramic must not include any oxides which are easily reduced to the metal, such as lead oxide. It is also important to note that the production of non-stoichiometric crystals resulting from heat-treatment of the glass-ceramic under reducing or, in some cases, neutral conditions is generally to be avoided. Non-stoichiometric titanium dioxide or titanate crystals, for example, could result in the deterioration of the electrical insulation characteristics.

3. *Crystallisation Characteristics of Glasses*

Perhaps the most important criterion in selecting a glass composition for glass-ceramic production is that it shall be capable of being crystallised without the use of prohibitively long heat-treatments. It is also important that the constituents of the glass and their proportions shall be chosen so that crystal types can be developed which will confer upon the final glass-ceramic the desired characteristics.

There are certain glasses which are extremely difficult, if not impossible, to devitrify even with prolonged heating at temperatures at which crystal growth might be expected. For example, potassium aluminosilicate glasses, especially the composition derived by melting potash felspar, $K_2O.Al_2O_3.6SiO_2$, are very difficult to crystallise. Fairly obviously, glasses of such high stability would be avoided as starting materials for glass-ceramics since, even if devitrification could be achieved by the use of a nucleation catalyst in the glass, the heat-treatment schedule would probably be so long as to be entirely uneconomic. On the other hand, compositions which devitrify so readily that crystal growth occurs during cooling and shaping of the glass are unsatisfactory, as we have seen. The requirements, on the one hand for stability during cooling and shaping of the glass, and on the other hand for rapid crystallisation during heat-treatment of the glass, tend to conflict. However, the use of a nucleation catalyst enables these conflicting demands to be reconciled.

Glasses which will crystallise reasonably easily during the reheating process will be those containing fairly high proportions of modifying oxides. This is because these oxides weaken the glass network structure by introducing non-bridging oxygen ions in place of bridging oxygen ions which link adjacent SiO_4 tetrahedra. As the proportion of non-bridging oxygen ions increases, the network structure becomes progressively weakened so that the atomic rearrangements necessary for crystallisation become increasingly probable. There will be upper limiting proportions of modifying oxides which can be present, governed by the requirement to avoid crystallisation of the glass while it is in its molten or semi-molten state.

Generally speaking, the usefulness of a particular modifying oxide in promoting crystallisation of the glass during the controlled heat-treatment process will be determined by the size and charge of the metallic cation. Small cations of high field strength will tend to be surrounded by a more ordered arrangement of oxygen ions than will large cations of low field strength. Thus with the small cations the transition to the ordered crystal structure can be accomplished more easily. On this basis, lithium oxide is likely to be a more useful constituent than sodium or potassium oxides and, of the alkaline earth oxides, magnesium oxide will perhaps be the most useful since the field strength of the cation is significantly higher than those of the calcium or barium ions. The latter ions, when present in high concentrations, however, would tend to promote devitrification and calcium or barium oxides could be used as major constituents of certain types of glass-ceramic. Zinc oxide also is a useful constituent of a glass-ceramic since the field strength of the zinc ion is fairly high so that this ion could exert a marked ordering effect on surrounding oxygen ions. This would increase the probability of devitrification for glasses containing high proportions of zinc oxide. The probable effect of zinc oxide upon the glass structure may be even greater than suggested by the simple concept of field strength which treats ions as if they were rigid spheres. The zinc ion, with 18 outer electrons, is not a noble gas cation and it therefore exerts a strong deforming influence upon surrounding anions. For example, the zinc ion in ZnO is surrounded by only four oxygen ions whereas the magnesium ion in MgO is surrounded by six oxygen ions. This is evidence of the greater deforming power of the zinc ion and this will be reflected in glass structure by a greater ordering effect of the zinc ion upon neighbouring oxygen ions.

4. *Glass Compositions for Glass-ceramic Production*

In chapter 2, discussing types of glass, reference was made to certain glass-forming systems which are important for the production of glass-ceramics. Actual glass-ceramic compositions are usually more complex then the ternary systems which were described and it is the purpose of this section to discuss practical examples of glass-ceramics derived from various oxide systems.

a. Glass-ceramics derived from the Li_2O–Al_2O_3–SiO_2 System. These compositions are of great importance, since photosensitively nucleated glass-ceramics are of this type, and other useful materials, including glass-ceramics having very low coefficients of thermal expansion are derived from this system.

In addition to the major constituents, the glass compositions also include a nucleation catalyst which may be a metal, titanium dioxide, a metallic phosphate or one of the other materials already described. Certain constituents of secondary importance are often included in the glasses to modify the melting and working properties or the crystallisation characteristics. These constituents may include alkaline earth oxides and alkali metal oxides. In some cases, a constituent which is present only in a small concentration can be extremely important because it exerts an apparently disproportionate effect upon the crystallisation behaviour. Potassium oxide is an example of such an oxide since, as we have seen, it promotes the formation of lithium metasilicate in preference to the disilicate in photosensitively nucleated glass-ceramics. In other cases it exerts a strong influence upon the form of crystalline silica which develops.

The types of glass-ceramic composition and the nature of the secondary constituent will become clear by reference to practical examples. Titania-nucleated glass-ceramics of the Li_2O–Al_2O_3–SiO_2 type having thermal expansion coefficients less than 15×10^{-7} and in some cases approaching zero have been produced. The compositions of these materials fall within the weight percentage range: SiO_2:53–75; TiO_2: 3–7; Li_2O: 2–15; Al_2O_3: 12–36.

Examples of these compositions are given in Table VII, a1 to a3. Similar glass-ceramics can also be prepared from glasses containing phosphorus pentoxide as a nucleation catalyst and compositions a4 and a5 are examples of these.

b. Glass-ceramics derived from the MgO–Al_2O_3–SiO_2 System. These glass-ceramics are important because they can be entirely free from alkali metal ions so that outstanding electrical characteristics, including low dielectric losses and high resistivities, can be attained. Small proportions of alkali metal oxides and other constituents are sometimes included in the glasses to modify the characteristics, however.

The major constituents of the glasses in weight percentages are in the range: SiO_2: 40–70; Al_2O_3: 9–35; MgO: 8–32. In addition, the glasses contain a nucleation catalyst such as TiO_2 (7 to 15 per cent) or P_2O_5 (0·5 to 6 per cent). Specific examples of compositions are given in Table VII, b1 to b6.

c. Glass-ceramics derived from the Li_2O–MgO–SiO_2 System. Metallic phosphates have been used to catalyse the crystallisation of glasses having weight percentage compositions in the range: SiO_2: 51–88; MgO: 2–27; Li_2O: 9–27 and P_2O_5: 0·5–6, together with various constituents of a secondary nature. Table VII gives examples of glass-ceramic compositions of this type (cl to c3).

Glass-ceramics of the Li_2O–MgO–SiO_2 type are of interest since in certain cases they have unusually high thermal expansion coefficients (up to 140×10^{-7}).

d. Glass-ceramics derived from the Li_2O–ZnO–SiO_2 System. Glass-ceramics possessing high mechanical strengths and other desirable properties can be produced from glasses of the lithium zinc silicate type. Suitable nucleation catalysts include metallic phosphates or metals such as copper, silver or gold. These glasses do not require irradiation in order to sensitise the metallic nucleation catalysts. The weight percentages of the major glass constituents lie in the range: SiO_2: 34–81; ZnO: 10–59; LiO_2O: 2–27, and these constituents should total at least 90 per cent of the glass composition. The nucleation catalysts include phosphorous pentoxide 0·5 to 6 per cent, gold 0·02 to 0·03 per cent, silver computed as AgCl 0·02 to 0·03 per cent or copper computed as Cu_2O 0·5 to 1 per cent. In addition to the essential constituents, alkali metal oxides (Na_2O, K_2O), alkaline earth oxides (MgO, CaO, BaO), aluminium oxide (Al_2O_3), boric oxide (B_2O_3) and lead oxide (PbO) can be present in the glasses in minor proportions.

Typical glass compositions are given in Table VII, d1 to d4, and these show that the proportions of the major constituents can be varied over a wide range. By selection of the composition, glass-ceramics having thermal expansion

TABLE VII

EXAMPLES OF GLASS-CERAMIC COMPOSITIONS

Glass reference	SiO_2	Al_2O_3	MgO	CaO	BaO	ZnO	PbO	CdO	Li_2O	K_2O	B_2O_3	TiO_2	P_2O_5	ZrO_2	F	Literature reference
a1	73·5	16·2	—	—	—	—	—	—	4·3	—	—	6·0	—	—	—	1
a2	67·8	20·8	—	—	1·0	1·0	—	—	3·9	—	—	5·5	—	—	—	1
a3	68·0	21·0	—	—	—	—	—	—	4·0	1·0	—	6·0	—	—	—	1
a4	63·6	19·8	2·8	2·7	—	—	—	—	8·1	—	—	—	3·0	—	—	3
a5	55·9	18·9	—	—	—	—	—	—	22·2	—	—	—	3·0	—	—	3
b1	42·8	30·2	14·0	—	—	—	—	—	—	—	—	13·0	—	—	—	2
b2	45·2	29·5	10·4	—	—	—	—	—	—	1·8	—	10·8	—	—	2·3	2
b3	46·7	28·9	13·3	—	—	—	—	—	0·9	—	—	10·2	—	—	—	2
b4	64·8	18·5	9·3	—	—	—	—	—	—	—	—	7·4	—	—	—	2
b5	52·3	20·3	24·4	—	—	—	—	—	—	—	—	—	3·0	—	—	3
b6	59·4	17·9	19·7	—	—	—	—	—	—	—	—	—	3·0	—	—	3
c1	73·1	—	7·0	—	—	—	—	—	10·5	—	6·4	—	3·0	—	—	3
c2	67·9	—	15·5	—	—	—	—	—	13·6	—	—	—	3·0	—	—	3
c3	64·7	7·7	11·7	—	—	—	—	—	12·9	—	—	—	3·0	—	—	3
d1	58·1	—	—	—	—	15·8	—	—	23·1	—	—	—	3·0	—	—	4
d2	48·2	—	—	—	—	39·2	—	—	9·6	—	—	—	3·0	—	—	4
d3	44·4	—	—	—	—	48·1	—	—	4·5	—	—	—	3·0	—	—	4
d4	37·2	—	—	—	—	50·5	—	—	9·3	—	—	—	3·0	—	—	4
d5	59·2	—	—	—	—	13·1	14·0	—	9·0	2·0	—	—	2·7	—	—	5
d6	47·8	—	—	—	—	11·0	29·9	—	7·3	1·6	—	—	2·4	—	—	6
e1	50·9	23·2	—	—	—	18·5	—	—	—	—	—	7·4	—	—	—	2
e2	49·0	19·6	—	—	—	29·4	—	—	—	—	—	2·0	—	—	—	2
e3	27·0	17·0	—	—	—	36·0	—	—	—	—	16·0	—	—	4·0	—	7
e4	—	6·0	—	—	—	50·0	—	—	—	—	40·0	—	—	4·0	—	7
e5	—	6·0	—	5·0	—	50·0	—	—	—	—	39·0	—	—	—	—	7
f1	53·4	17·7	—	17·8	—	—	—	—	—	—	—	11·1	—	—	—	2
f2	66·7	15·5	—	6·7	—	—	—	—	—	—	—	11·1	—	—	—	2
f3	48·2	17·6	—	—	21·9	—	—	—	—	—	—	12·3	—	—	—	2
f4	39·5	21·9	—	—	26·3	—	—	—	—	—	—	12·3	—	—	—	2
f5	31·1	15·5	—	—	—	—	42·3	—	—	—	—	11·1	—	—	—	2
f6	15·6	4·4	—	—	—	—	68·9	—	—	—	—	11·1	—	—	—	2
f7	27·3	27·3	—	—	—	—	—	36·3	—	—	—	9·1	—	—	—	2
f8	44·7	22·3	—	—	—	—	—	22·3	—	—	—	10·7	—	—	—	2

References: 1. Stookey (1959b), 2. Stookey (1960), 3. McMillan and Partridge (1963a), 4. McMillan and Partridge (1963b), 5. McMillan and Hodgson (1966), 6. McMillan and Hodgson (1967), 7. McMillan and Partridge (1969).

coefficients within the range 43×10^{-7} to 174×10^{-7} can be produced, and this is evidence of the versatility of these materials.

In later developments, it was shown that lead oxide, PbO, could be included as a major constituent in glass-ceramics of this general type. Up to 30 per cent PbO can be incorporated, as illustrated by compositions d5 and d6. Materials of this type also possess high thermal expansion coefficients and the high lead oxide content confers improved dielectric loss characteristics.

e. Alkali-free glass-ceramics having high ZnO contents. Glass-ceramics derived from the $ZnO–Al_2O_3–SiO_2$ system which contain no alkali oxides are possible as exemplified by compositions e1 and e2 in Table VII. Partial replacement of SiO_2 by B_2O_3 as in composition e3 or complete replacement as in compositions e4 and e5 gives glass-ceramics which have flow characteristics that make them suitable for special coating and sealing processes.

f. Glass-ceramics derived from various alumino-silicate systems. In addition to the glass-ceramics systems discussed above, compositions containing major proportions of various other oxides have been developed. Compositions f1 to f8 illustrate some of the possibilities but this list is by no means exhaustive.

The examples of glass-ceramic compositions quoted will have shown that the process of controlled crystallisation can be applied to a wide variety of glasses. Since each type of glass-ceramic contains characteristic crystal phases, materials having widely different physical properties can be produced. This versatility represents one of the most valuable features of the glass-ceramic process.

Chapter 4

THE GLASS-CERAMIC PROCESS

The glass-ceramic process comprises the preparation of a homogeneous glass, the shaping of the glass to produce the required articles and, finally, the application of a controlled heat-treatment process to convert the glass into a microcrystalline glass-ceramic.

A. The Preparation of Glasses

Glasses are made by heating together a mixture of raw materials (known as "batch") at a sufficiently high temperature to permit the materials to react with one another and to encourage the escape of gas bubbles from the melt; this latter process is referred to as refining the glass. Upon completion of the refining process, the glass is cooled from the melting temperature to the working temperature where the glass has a higher viscosity. Various shaping methods can then be applied to the glass to produce articles of the required form.

1. *Raw Materials of Glass Manufacture*

In selecting raw materials the most important aspect to be taken into account is the purity. More specifically, the composition of the raw material must be constant within fairly close limits so that proper control of the glass composition can be exercised. Economic considerations will also play an important part in the choice of materials so that while for the production of expensive types of glasses, such as optical glass, materials of very high purity will be employed, less expensive and somewhat less pure materials will be used for the cheaper types of glasses. In the case of glass-ceramics, materials of quite high purity will be used since some types of impurity, even in quite small concentrations, could affect the crystallisation characteristics of the glass.

Undoubtedly, the most important constituent of the glass is silica, since it is the principal glass-forming oxide present in most glasses. The chief source of silica in the glass batch is quartz sand and the best sands will have SiO_2 contents of 99·5 per cent or higher. Small amounts of alumina, Al_2O_3, may be present and iron oxide, Fe_2O_3, will certainly be present. Although this oxide confers a greenish colour to glass, its presence is not apparent in the opaque glass-ceramic. The alkaline earth oxides such as CaO and MgO are often introduced in the form of their naturally occurring carbonates – limestone in the case of CaO and magnesite in the case of MgO. A dolomite limestone which contains both calcium and magnesium carbonates may also be used if the glass composition includes both oxides. Barium carbonate is a convenient source of barium oxide. Aluminium oxide is often introduced in the form of hydrated alumina, $Al_2O_3.3H_2O$, but calcined alumina may also be used. Felspars which have the general formula $R_2O.Al_2O_3.6SiO_2$ also form a source of alumina, either as potash felspars, soda felspars or mixed felspars. Of course, the felspars also introduce silica and alkali oxides in addition to their alumina content. A more usual source of sodium oxide, however, is soda ash, Na_2CO_3, and potassium oxide is often introduced as the carbonate K_2CO_3. Part of the alkali content of a glass may be introduced in the form of a nitrate. This is useful since these compounds assist melting and refining of the glass and also act as oxidising agents where these are required. Lithium oxide can be conveniently introduced as the carbonate Li_2CO_3, although economic considerations may dictate the use of a lithium-bearing mineral such as petalite or spodumene; these materials, in addition, introduce silica and alumina and the achievement of the desired glass composition may not always permit their use. The usual sources of boric oxide are boric acid, $B_2O_3.3H_2O$ or borax $Na_2B_4O_7.10H_2O$. Small amounts of arsenic or antimony oxides may be introduced as refining agents and cerium dioxide is often used as an oxidising agent. In some glasses this oxide also exerts a refining effect.

2. *The Melting of Glasses*

The raw materials are accurately weighed out and after thorough mixing are charged into a furnace maintained at the melting temperature which can range from 1250 to 1600°C, depending upon the glass composition. Scrap glass of the correct composition, known as cullet, is often included with the batch since this assists in melting and homogenising the glass. The glass can be melted in crucibles or pots, but for large-scale production melting is carried out in continuous furnaces in which a bath of molten glass is continually replenished by charging batch into one end of the furnace while molten glass is removed for shaping at the other end of the furnace. Furnaces of this type are known as tank furnaces and they may be heated by gas or oil flames or, in recent practice, by passing a heavy electric current through the molten glass. High-grade refractory materials are used for the construction of the container for the molten glass to minimise changes of composition of the glass due to its corrosive action upon the refractory materials. The refractories used include high-grade fireclay or sillimanite-type refractories consisting principally of mullite. In addition, electrocast refractories based on mullite–zircon compositions are often used. In special cases a platinum-lined tank furnace may be utilised. This would be the case where glass of exceptional homogeneity and closely controlled properties is required; this procedure is adopted in the manufacture of glass-ceramics for missile radomes where a material which is very homogeneous dielectrically is needed.

The high temperatures in the glass-melting furnace ensure that vigorous reactions take place. The alkali carbonates react with the silica with the evolution of carbon dioxide, alkaline earth carbonates break down and the oxides are taken into solution and hydrates or nitrates also present break down, permitting the resulting oxides to combine with the siliceous melt. Violent agitation of the molten mass occurs due to the escaping gases and this assists mixing and maintains the reactions at a high rate. Eventually, complete solution and fusion of the materials is effected and there remains the process of refining or freeing the melt from bubbles. The rate of rise of a bubble is proportional to the square of its diameter so that large bubbles rise much more rapidly than fine ones. For this reason refining agents are included in the glass batch, the function of these being to evolve gas at late stages during the melting process and thus to produce large bubbles which sweep the melt clear of fine bubbles ("seed"). Refining agents include arsenic and antimony trioxides. Some refining occurs due to gas bubbles re-dissolving in the melt when it is cooled, and glass may contain appreciable amounts of dissolved gases such as carbon dioxide, oxygen and sulphur dioxide.

The refining stage is followed by cooling the glass to its working temperature, which may be several hundred degrees below the refining temperature. In tank furnaces this is accomplished by causing the glass to flow into a separate cooler chamber of the furnace. Here the glass will be brought to as uniform temperature as possible so that it will possess a uniform viscosity.

3. *Shaping Processes*

Perhaps the simplest shaping operation available to the glass maker is that of casting. Because of its simplicity, gravity casting has limited applicability and is restricted to the production of plates or shallow rings and the like. Centrifugal casting, in which molten glass is introduced into a mould rotating at high speed so that the glass is forced to assume the internal shape of the mould, is of greater value and can be used to produce shapes which are difficult to make by other means. Conical shaped radomes for missiles are made by this method, for example. Casting processes are useful for glasses having short working ranges, such as alkali-free glasses. Flat plates of glass can be produced by rolling, in which a continuous stream of glass from a tank furnace passes between water-cooled rolls. Glass in sheet form is also made by a continuous-drawing process in which a ribbon of glass is drawn from a pool of molten glass in a furnace. Rod and tubing are made by a drawing process; in this case, molten glass flows onto a rotating mandrel and is drawn from the tip of the mandrel in the form of a continuous rod or tube. Pressing of glass, employing suitable moulding equipment, is used for the production of lens- or dish-shaped articles, and fully automatic presses which attain high production rates are available. Many types of hollow-ware are made by blowing and, with fully automatic machines, extremely high rates of production are possible.

This brief outline of glass-shaping processes will serve to show that glass as a material can be manufactured by mass production processes and that it can be produced in a wide variety of forms. Many glass-ceramic compositions can be shaped by all of the available glass-working techniques, so that the versatility of glass is retained in the new materials.

During the shaping of glass, internal stresses are produced due to the presence of temperature gradients within the glass during cooling, and these stresses must be removed by annealing or they may result in fracture of the glass-ware. Annealing is achieved by bringing the glass to a uniform temperature within the "annealing range" where its viscosity is about 10^{12} to 10^{14} poises. The stresses are relaxed due to viscous flow and the glass is then cooled at a sufficiently slow rate to prevent the establishment of large temperature gradients within it. The cooling rates which can be employed are determined by the thickness of the glassware being annealed. For thin glass the whole process may be completed in an hour or so but thick heavy pieces may require several hours or even days to ensure that any residual stresses are kept within safe limits. Annealing can be accomplished in intermittent kilns but more often it is carried out in continuous furnaces.

B. Conversion of the Glass to a Microcrystalline Ceramic

1. *The Heat-treatment Process*

a. Nucleation and crystallisation processes. The object of the heat-treatment process is to convert the glass into a microcrystalline ceramic having properties superior to those of the original glass. It is especially important to achieve a high mechanical strength and since this is favoured by a fine-grained microstructure, the aim is to produce a glass-ceramic containing crystals of small dimensions which are closely interlocked. The production of large numbers of small crystals rather than a smaller number of relatively coarse crystals poses the requirement for efficient nucleation and this in turn means that careful control must be exercised over the nucleation stage of the heat-treatment. Having nucleated the glass, it is necessary to raise the temperature further in order to permit crystal growth upon the nuclei. The rate of temperature rise will also be carefully controlled since it will be necessary to avoid deformation of the glass-ceramic. If the rate of heating is too high, the rate of crystal growth may not be sufficiently rapid to ensure that a rigid crystalline "skeleton" is present at all temperatures. With slower heating, deformation should not occur since, although there will be a high proportion of glass phase present, this will diminish progressively in amount as the temperature is raised, being replaced by more refractory crystalline phases. Throughout this heating stage the composition of the glass phase will be changing progressively as various crystals are precipitated, and in many cases the effect of the crystallisation will be to increase the refractoriness of the residual glass phase. In addition to its undesirable effect of causing deformation, rapid heating is also to be avoided since it may result in cracking of the glass-ceramic. This danger arises because some of the crystals formed have different densities from the glass phase and the volume change which accompanies crystallisation can result in the generation of stresses in both the glass phase and the crystal phases. With slow heating these stresses are relieved by viscous flow of the glass phase. The slow increase of temperature will continue until an upper limiting temperature, referred to as the upper crystallisation temperature is reached. This temperature is one at which crystallisation will proceed rapidly without leading to deformation of the glass-ceramic due to softening of the residual glass phase or to melting of the least refractory crystalline phase. By maintaining the glass-ceramic at the upper crystallisation temperature for a suitable period, almost complete crystallisation can be achieved so that only a very small proportion of a residual glass phase will be present.

A clear idea of the various stages of the heat-treatment process will be obtained by reference to Fig. 30 which represents an idealised heat-treatment schedule for a glass-ceramic. It is proposed to consider the different parts of the heat-treatment schedule in further detail so that the various factors of importance may be emphasised.

The first stage of the process involves heating the glass from room temperature to the nucleation temperature. Generally speaking, the rate of heating employed here is not critical so far as the crystallisation process is concerned. It is limited mainly by the requirement that dangerously high stresses which might cause cracking shall not be generated due to temperature gradients within the glass articles. The thickness of the glass-ware will chiefly determine the rate which can be employed, although the thermal expansion coefficient of the glass will also play a part since, of course, glasses with low expansion coefficients can withstand higher temperature gradients without cracking than can the glasses with high expansion coefficients. Normally heating rates between 2°C and 5°C per minute will be employed, although for thin glass-ware rates as high as 10°C per minute can safely be used.

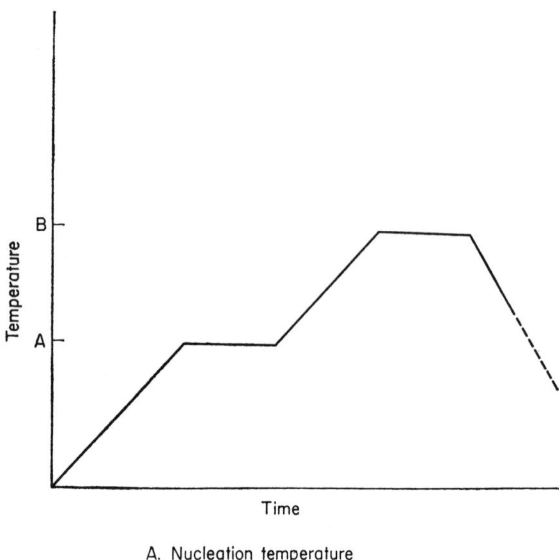

FIG. 30. Idealised heat-treatment schedule for a glass-ceramic.

The optimum nucleation temperature generally seems to lie within the range of temperatures corresponding with viscosities of 10^{11} to 10^{12} poises. The temperature within this range which gives optimum nucleation is determined by experimentation. As a first approximation, the optimum nucleation temperature lies between the Mg point and a temperature 50°C higher than this. A more exact determination of the optimum nucleation temperature may be carried out by the following method. A droplet of glass is melted in a miniature furnace provided with means for measuring the temperature of the glass droplet accurately and rapidly, and arranged so that the droplet can be under microscopic observation. The droplet is cooled to an arbitrarily chosen temperature, held at this temperature for a minute or so and then reheated to the lowest liquidus temperature of the glass. (When several different crystal phases can be produced by the devitrification of a glass it will possess a corresponding number of liquidus temperatures.) If on reheating to the lowest liquidus temperature no crystallisation occurs, the glass droplet is completely remelted and cooled to a slightly lower temperature than before and is again reheated to observe crystal formation, if any. This procedure is continued until the temperature of maximum nucleation has been determined. The period of time for which the glass is maintained at the nucleation temperature will usually be from 0·5 to 2 hours, although longer periods may not have a detrimental effect. Although nucleation can occur at any temperature between the temperature of optimum nucleation and the annealing point, the use of lower temperatures for nucleation can greatly increase the time for nucleation to be completed since the rapid increase of viscosity as the temperature falls leads to great reduction of the nucleation rate. For example, at the annealing point corresponding to a viscosity of $10^{13 \cdot 3}$ poises, a period of 100 hours may be required to achieve satisfactory nucleation of glasses containing titanium dioxide as the nucleating agent. An important point to note in connection with the nucleation process is that to achieve a glass-ceramic of optimum strength, the glass must be cooled below the maximum nucleation temperature before the nucleation heating stage. This will often occur in the natural course of events, since the material will be cooled in the glassy state and annealed at a temperature below the nucleation temperature, followed by cooling to room temperature. In some cases, however, the crystallisation heat-treatment process may follow immediately from the shaping of the glass-ware and in these cases it is necessary to cool the glass substantially below the nucleation temperature, and in some cases the temperature interval necessary may be as much as 100°C to 300°C.

As an alternative to the procedure outlined in the preceding paragraph, the optimum temperature for nucleation may be determined by heat-treating specimens for a fixed period (eg 1 hour) at various temperatures in the nucleating range and then transferring them to a furnace held at a crystal growth temperature. After a predetermined time the specimens are removed from the furnace, effectively quenching them and arresting the growth process. By means of optical and electron microscopy, the number of growth centres and mean crystal diameters can be established and shown to be

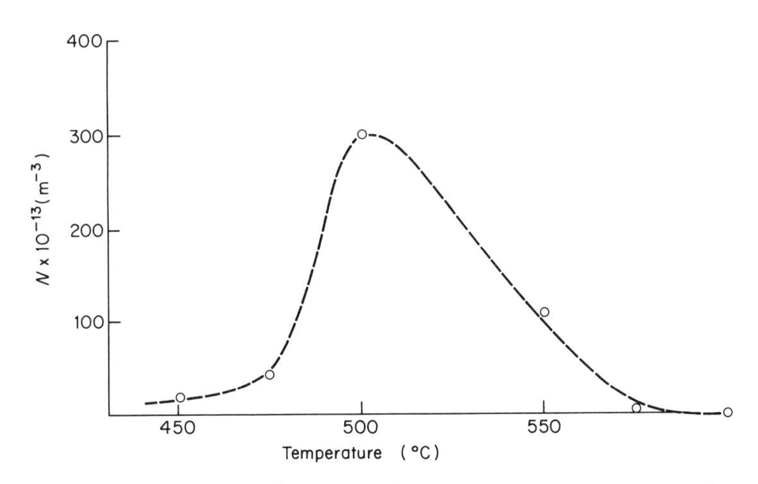

FIG. 31. General form of nucleation density, N, versus temperature curve for a glass of the molecular percentage composition SiO_2 69; Li_2O 30; P_2O_5 1.

strongly dependent upon the nucleation temperature used. Figure 31 summarises the results of such a study by Harper *et al.* (1970) on a glass-ceramic of the molecular percentage composition Li_2O 30; SiO_2 69; P_2O_5 1. For this composition there is a clearly defined optimum nucleation temperature of about 500°C. In the same investigation, the value of P_2O_5 as a nucleating agent was also demonstrated. Figure 32 shows micrographs of specimens of two glasses nucleated for 1 hour at 500°C followed by heat-treatment at 750°C for 1 hour to accomplish crystal growth. Quite clearly, the glass containing P_2O_5 developed a much finer-grained microstructure. For this glass the nucleation density was 3×10^{15} m^{-3} as compared with 1×10^{13} m^{-3} for the P_2O_5 free glass; the corresponding mean grain sizes were 5 μm and 60 μm. For other glasses in the Li_2O–SiO_2–P_2O_5 system Harper and McMillan (1972) reported even higher nucleation densities of the order of 10^{20} m^{-3}.

In the foregoing, the determination of nucleation densities was by an "indirect" method. That is, it was necessary to grow crystals on the nuclei in order to render them amenable to evaluation by optical or electron microscopy. Using high resolution transmission electron microscopy, however, it is possible to omit the growth stage and to determine nucleation densities for specimens subjected only to the nucleation treatment. In this case the crystals to be resolved are very small indeed being in the range 5 to 20 nm. Hing and McMillan (1973a) were successful in studying nucleation in a glass of the molecular composition Li_2O_3 30; SiO_2 69; P_2O_5 1 in this way. In agreement with Harper *et al.* (1970) they concluded that the optimum nucleation temperature was 500°C but the nucleation density at 500°C was higher by a factor of 10^6 than that reported by the previous workers. It seems likely that the reason for this discrepancy is that only a fraction of the crystalline particles developed at the nucleating temperature are effective in promoting crystal growth. The disappearance of nuclei as a result of a coarsening process during the heat-treatment is probably responsible for this effect.

The elimination of nuclei by a coarsening mechanism was demonstrated by McMillan (1974) who determined the effects of duration of nucleation treatment at 550°C for a Li_2O–SiO_2–P_2O_5 glass. The results, summarised in Fig. 33, indicate that there is an optimum duration of heat-treatment in order to produce a glass-ceramic having the finest-grained microstructure. The micrographs in Fig. 34 show the marked effects on microstructure of varying the duration of nucleation heat-treatment. The reduction in nucleation density after the optimum nucleation time was consistent with the occurrence of a diffusion controlled coarsening process.

Following the nucleation stage, the temperature of the glass is increased at a controlled rate sufficiently slowly to permit crystal growth to occur so that

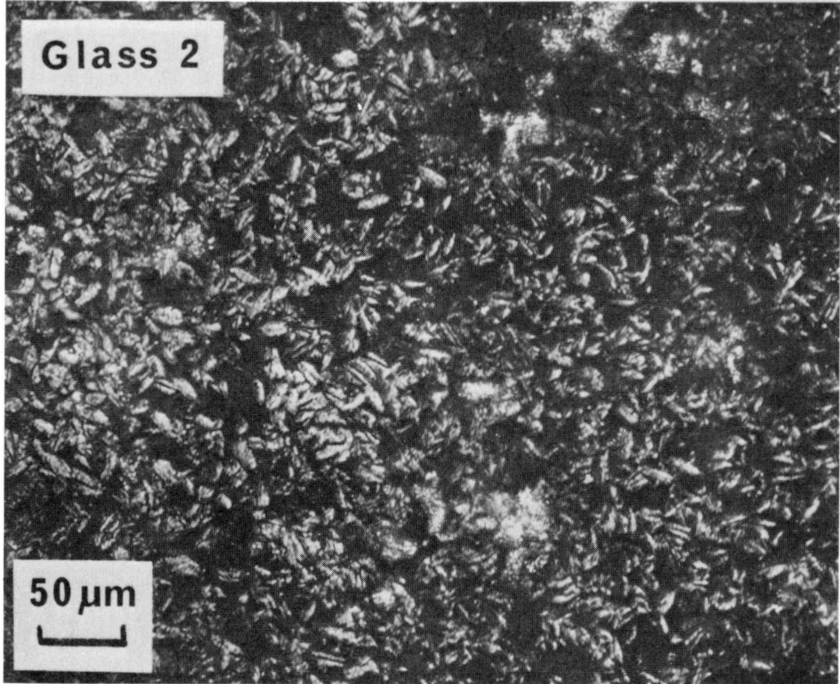

FIG. 32. Microstructures of glass-ceramics. Molecular percentage compositions: Glass 1 SiO_2 70; LiO_2 30 Glass 2 SiO_2 69; Li_1O 30; P_2O_5 1.

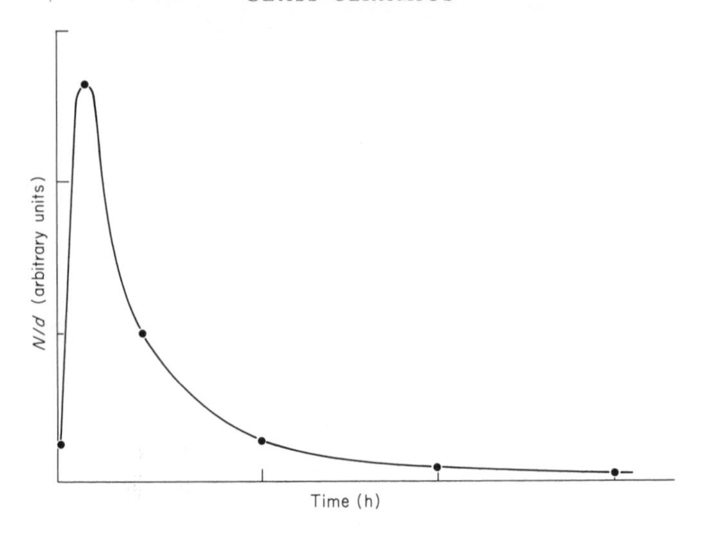

FIG. 33. The ratio of nucleation density to mean crystal diameter (*N*/*d*) as a function of nucleation time for SiO_2 69; Li_2O 30; P_2O_5 1 (mol. per cent) glass. An improved microstructure is indicated by an increase in the value of the ratio.

FIG. 34. Effects of duration of nucleation treatment on the microstructure of a glass-ceramic (composition as for Fig. 33). Nucleation treatments: (a) 550°C/2 min, (b) 550°C/20 min, (c) 550°C/1 hr, (d) 550°C/3 hr. (× 144)

deformation of the glass article will not take place. Crystallisation occurs increasingly rapidly as the liquidus temperature of a predominant crystalline phase is approached but, to prevent deformation in the early stages when the glass phase predominates, a heating rate not usually exceeding 5°C per minute is employed. The permissible heating rate can readily be determined by experiments in which rods supported on two knife edges are subjected to the nucleation heat-treatment stage and are then heated at various rates. The sag occurring at the centre of the rod is afterwards measured and used as a guide to decide upon an acceptable rate of heating.

The upper crystallisation temperature for a glass-ceramic is chosen so that maximum crystallisation can be achieved without leading to excessive deformation of the material. The temperature at which the final product will deform appreciably will correspond to the liquidus temperature of the predominant crystal phase since increasing the temperature above this liquidus will cause the phase to redissolve. The upper crystallisation temperature will be lower than the temperature at which the predominant crystalline phase will redissolve by a suitable interval, usually 25 to 50°C.

One method of determining the liquidus temperature of the various crystalline phases present is to heat a previously nucleated rod of glass in a gradient furnace in which the temperatures throughout the length of the furnace are measured accurately. After this treatment, it is observed that the portion of the rod which was heated above the nucleation temperature but below the highest liquidus temperature will contain crystalline material. The region of the rod which was heated above the highest liquidus temperature, on the other hand, will be clear and free from crystals. Within the crystalline part of the rod various crystalline phases will occur in different zones and the nature of the phases present in any particular zone will depend on the temperature to which it was heated. X-ray diffraction analysis and microscopic examination may be used to identify the phases present in the various zones.

A convenient method for determining the highest liquidus temperature only, is to nucleate a droplet of glass in the miniature furnace as described earlier, to reheat it to cause crystallisation and to heat it further to the temperature at which all of the crystals are redissolved.

The upper crystallisation temperature, determined by one of the foregoing methods, is maintained for a period of at least one hour, but longer holding periods may be employed if this is necessary in order to achieve the desired degree of crystallinity in the glass-ceramic. After completion of the holding period the glass-ceramic is cooled to room temperature. Cooling can be quite rapid since the glass-ceramics can withstand fairly high temperature gradients because of their high mechanical strengths. A certain amount of care is exercised in cooling materials of high thermal expansion coefficient although, even for these, cooling rates as high as 10°C per minute can often be employed and for the low expansion materials faster cooling still is quite safe. Unlike glasses, glass-ceramics do not require annealing to prevent the generation of permanent strains in the materials.

Some of the crystalline phases present may exhibit structural changes in certain temperature zones and these changes are accompanied by alterations of density and thermal expansion coefficient. In a conventional ceramic, structural changes of this type can cause breakage unless the ceramic is cooled rather slowly through the critical temperature zone. For glass-ceramics, this limitation is not found to the same degree since quite rapid cooling through critical temperature zones does not cause fracture of the material. The superiority of glass-ceramics in this respect may be due to the very small sizes of the crystals present, since the stresses generated as a result of volume changes would be small.

b. Changes in physical characteristics brought about by the heat-treatment process. In this section it is proposed to compare briefly the properties of glass-ceramics with those of the glasses from which they are derived in order to illustrate the changes in physical characteristics, most of which are beneficial, which can be achieved by controlled devitrification. In chapter 5 the properties of glass-ceramics will be studied in much greater detail.

An obvious change brought about by the heat-treatment is the conversion of the transparent glass to an opaque polycrystalline material. The opacity of the glass-ceramic is due to scattering of light at interfaces between adjacent crystals and between crystals and the residual glass phase due to the differences in refractive indices of the phases. In certain instances, where the crystals are small and the refractive indices of the various phases are fairly closely matched, the glass-ceramic may be transparent or translucent, however.

In addition to the obvious change in appearance, there is another, more subtle change which is only apparent when the materials are examined under high magnifications. To the eye and to the touch glass-ceramics appear perfectly smooth, but examination by the electron microscope reveals that their surfaces are not so smooth as those of the parent glasses. Figures 35 and 36 show the appearance of a typical glass and the corresponding glass-ceramic. Clearly, the result of heat-treatment is to produce a generally undulating surface made up of rounded crystal boundaries and occasionally there are angular crystals which project above the mean surface level. Even though the surface of a glass-ceramic is less smooth than that of a glass, this difference is not significant in most practical applications. A further point is that glass-ceramic surfaces will be considerably smoother than those of unglazed conventional ceramics.

FIG. 35. The surface of a parent glass before heat-treatment (× 10 000).

FIG. 36. The surface of a glass-ceramic (× 10 000).

The specific gravity of a glass-ceramic is very often different from that of the parent glass because small volume changes may occur during the heat-treatment process. These changes may involve either a slight contraction or a slight expansion of the material but they would not usually exceed a 3 per cent volume change. This may be contrasted with the large volume changes which occur during the drying and firing of conventional ceramics which may quite often total as high as 40 to 50 per cent. The volume changes in glass-ceramics are a result of the overall differences in specific gravity of the crystalline phases which are formed as compared with those of the parent glasses. The crystal phases formed may have higher or lower densities than the glass so that the net effect of crystallisation can be to cause a contraction or an expansion of the material. Obviously there will be limited ranges of composition where the net volume change could be zero since the contraction resulting from the production of one crystal phase may be exactly balanced by the expansion due to the production of another phase. Some idea of the volume changes which can occur in practice is given by the figures for densities of glasses and glass-ceramics in Table VIII. The value for the titania nucleated ZnO–Al_2O_3–SiO_2 glasses are especially interesting since they illustrate the possibilities of a positive, zero or negative volume change during heat-treatment.

The relatively small dimensional change which takes place during heat-treatment constitutes one of the advantages of the glass-ceramic process over conventional ceramic manufacturing techniques, since it enables articles to be produced to much closer dimensional tolerances.

The thermal expansion coefficients of glass-ceramics are generally markedly different from those of the parent glasses. Devitrification of the glass may result in raising or lowering of the thermal expansion coefficient depending upon the types of crystal which are formed. The formation of cristobalite, for example, will give materials with high expansion coefficients, especially for the temperature range 20°C to about 200°C. On the other hand, the formation of lithium aluminosilicate type crystals such as beta-spodumene or beta-eucryptite can give materials having very low thermal expansion coefficients. The thermal expansion curves for glasses and glass-ceramics given in Fig. 37 show that parent glasses having very similar thermal expansion coefficients can give rise to glass-ceramics having markedly different coefficients of expansion due to the formation of different crystal phases.

Another change in physical characteristics brought about by devitrification is the increase in refractoriness of the material.

Although there are a number of ways of defining this characteristic, a convenient method of comparing materials of the type under consideration is to make use of the dilatometric softening temperature. Providing the thermal expansion tests are carried out under similar conditions, this temperature

TABLE VIII

DENSITIES OF GLASSES AND CORRESPONDING GLASS-CERAMICS*

Glass composition – weight per cent										Density of parent glass (g/cm^3)	Density of glass-ceramic (g/cm^3)
SiO_2	B_2O_3	Al_2O_3	ZnO	MgO	BaO	CdO	Li_2O	TiO_2	ZrO_2		
57·6	4·7	15·2	—	—	—	—	5·2	12·1	3·6	2·52	2·57
48·5	—	14·6	34·0	—	—	—	—	2·9	—	3·17	3·13
41·7	—	9·3	41·6	—	—	—	—	7·4	—	3·23	3·23
50·9	—	23·2	18·5	—	—	—	—	7·4	—	2·92	2·99
48·5	—	21·3	—	4·8	—	15·0	—	10·1	—	2·87	2·86
46·5	—	24·1	—	—	17·6	—	—	12·3	—	2·96	2·96

* Data extracted from British Patent no. 829 447 (Stookey, 1960).

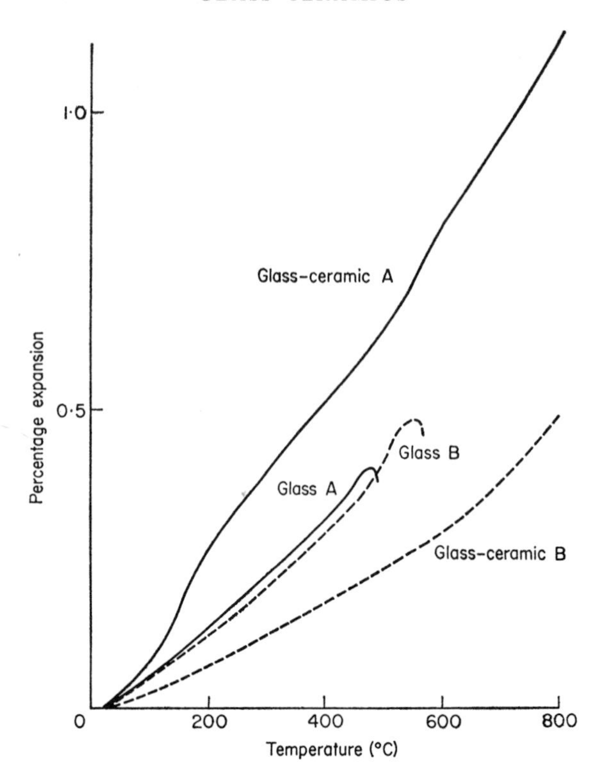

FIG. 37. Thermal expansion curves of glass-ceramics and parent glasses.

provides a reasonable basis for comparison. The data given in Table IX for glass-ceramics and their parent glasses show that the crystallisation process increases the softening temperatures very considerably.

Perhaps the most striking and important change in characteristics which is brought about by the crystallisation heat-treatment is the increase of mechanical strength. The cross-breaking strength or modulus of rupture of most glasses is in the vicinity of 70 MNm^{-2} (~ 10 000 lb in^{-2}) for rods of ~ 5 mm in diameter. For glass-ceramics, however, values of 200 to 280 MNm^{-2} (~ 30 000 to 40 000 lb in^{-2}) are generally obtained and for some materials even higher mean strengths are possible.

Generally speaking, the electrical properties of glass-ceramics are superior to those of the parent glasses and in particular the electrical resistivities are higher and the dielectric losses are lower. This improvement in properties can to a large extent be attributed to the tighter binding of ions, especially alkali metal ions within the regular crystal lattices as compared with the disordered glass structure.

TABLE IX

THERMAL EXPANSION COEFFICIENTS AND DILATOMETRIC SOFTENING TEMPERATURES OF GLASSES AND CORRESPONDING GLASS-CERAMICS

Weight percentage composition*					Thermal expansion coefficients × 10^7 (20–400°C)		Dilatometric softening temperatures (°C)	
SiO_2	Li_2O	MgO	Al_2O_3	K_2O	Glass	Glass-ceramic	Glass	Glass-ceramic
81·0	12·5	—	4·0	2·5	84·3	113·0	540	910
77·5	12·5	—	10·0	—	78·0	50·0	530	930
62·1	1·9	17·6	18·4	—	42·0	50·3	790	1000
66·4	10·0	3·0	20·6	—	63·4	0·7	640	1000
60·2	8·5	2·8	28·5	—	60·5	−42·4	720	1000
76·7	14·7	8·6	—	—	92·8	92·5	525	890
62·4	12·4	25·2	—	—	78·8	74·8	600	930

* N.B. The glasses also contain a small proportion of P_2O_5 to act as a nucleation catalyst.

2. *Methods of Studying the Heat-Treatment Process*

The heat-treatment process is a critical stage in the production of a glass-ceramic and must be carefully controlled to ensure that the desired types and proportions of crystals are formed. It is necessary, therefore, to make use of sensitive techniques for studying the crystallisation processes in order to ensure that the optimum heat-treatment schedule is applied. The methods used are all physical ones since, of course, no change in the chemical composition occurs during the conversion from the glass to the ceramic state.

a. X-ray diffraction analysis. As is well known, crystalline substances give sharp X-ray diffraction spectra which can be photographed and used as a positive means of identifying crystals by comparison with standard data such as the A.S.T.M. Index. Glasses, on the other hand, show rather diffuse X-ray diffraction patterns with complete absence of sharp lines. X-ray diffraction analysis, therefore, provides an excellent means for investigating the crystallisation of glasses.

The technique can be applied in a number of ways, all of which are useful but all of which have their limitations. An obvious application of the method is in the identification of the crystalline phases in the fully crystallised glass-ceramics. X-ray diffraction data exist for the various crystalline forms of silica such as quartz or cristobalite, for crystalline silicates such as lithium or zinc silicates, and for alumino silicate crystals such as betaspodumene or cordierite. Quite often, however, X-ray diffraction patterns will be obtained for hitherto unknown crystal types and in such cases a lengthy procedure of synthesising the unknown crystal, of carrying out X-ray diffraction studies and of working out the crystal structure will be necessary. A further possible complication is that the X-ray diffraction pattern observed may correspond fairly closely to that of a known crystal but may not be absolutely identical. Possible reasons for this are that the crystals produced by the devitrification of a glass may not be completely pure since certain ions in the crystal structure may be partially replaced by other ions of similar size and charge. Also, some crystals form solid solutions rather easily. For example, beta-spodumene and quartz can give rise to a whole series of solid solutions with consequent modifications of the X-ray diffraction spectra.

Merely identifying the crystal phase in the final glass-ceramic does not give the most useful information. It is of greater value to determine the sequence in which various crystals arise during the heat-treatment process and to define the conditions of temperature and time under which the various phases are formed. For this purpose, specimens of the glass are subjected to the heat treatment schedule and are removed individually after pre-determined time intervals. The specimens are cooled quickly to "freeze in" the conditions obtaining at the instant they are removed from the heat-treatment furnace and are then subjected to X-ray diffraction analysis. By this means, the complete crystallisation process from the nucleation stage to the final crystallisation stage can be followed and the separation of different crystal phases can be correlated with the time-temperature schedule of the heat-treatment process. The method has its limitations, however, since it depends on the use of quenched samples which gives a series of "snapshots" rather than a continuous record. An improved technique is to use a high temperature X-ray camera in conjunction with a recording diffractometer so that the gradual intensification of lines in the X-ray spectrum corresponding with the development of particular phases can be continuously monitored throughout the heat-treatment schedule. The limitation of this technique is that the development of only one crystalline phase can be followed easily, although scanning of the spectrum might under ideal conditions allow the growth of more than one phase to be studied.

In addition to identifying the crystal phases and revealing the conditions under which they are formed, X-ray diffraction analysis can be used to give an approximate idea of the amounts of various phases which are present in a glass-ceramic. This method can only be applied for those crystals which have been positively identified. The technique involves the preparation of a series of "synthetic" glass-ceramics which are powdered mixtures of the known crystal phase and of a glass in various proportions. It is wise to use the parent glass of the glass-ceramic in these mixtures so that X-ray scattering effects will be similar in the synthetic mixture and in the actual glass-ceramic. X-ray diffraction spectra are obtained of the crystal-glass mixtures and of the glass-ceramic using identical conditions of exposure and processing of the X-ray film. By carrying out photometric observations on strong lines in the diffraction photographs for the synthetic mixtures, a calibration curve can be drawn relating the intensity of a particular diffraction line with the percentage of the crystal phase. This curve can then be used to arrive at the proportion of crystal phase in the glass-ceramic from similar measurements on X-ray diffraction photographs for the glass-ceramic. The same procedure is carried out separately for each known phase in the glass-ceramic.

b. Optical and electron microscopy. For most glass-ceramics, the structural features of interest are often below the limit of resolution of the optical microscope. This is especially true for nucleating particles which can have dimensions less than 10 nm and even in the fully developed glass-ceramic microstructure, the mean crystal size is frequently only of the order of 1 micron. For this reason, electron microscopy has generally proved more useful for glass-ceramic microstructural studies. There are cases, however, where optical microscopy of polished and etched sections can yield valuable information.

The use of optical microscopy to investigate the influence of compositional changes on glass-ceramic microstructures was illustrated by the work of Harper and McMillan (1972). A number of compositions basically of the Li_2O–SiO_2 type were studied, some containing a small proportion of P_2O_5 as a nucleating agent and others free from this oxide. As shown in Fig. 38, depending on the composition, microstructures ranging from those containing coarse spherulitic crystals to others containing relatively small crystals were observed. The inclusion of P_2O_5 was shown to promote the desirable fine-grained microstructure.

Where higher magnification than can be achieved optically is required, the scanning electron microscope (SEM) is of great value. Though it has a lower limit of resolution of about 20 nm this equipment allows many important structural features of a glass-ceramic to be elucidated. This technique has the advantage that specimen preparation is relatively simple and allows examination of fractured surfaces as well as of polished and lightly etched surfaces. It is in fact often found that a light etch in dilute ($\sim$2 per cent) hydrofluoric acid is advantageous in enhancing the microstructural detail that can be observed. Generally, the satisfactory examination of glass-ceramics in the SEM requires the specimens to be precoated with an evaporated electrically conducting film (eg gold-palladium). Otherwise, charging of the

FIG. 38. Microstructures of Li_2O–SiO_2–P_2O_5 glass-ceramics. (The compositions of the glass-ceramics are given in Fig. 16.)

specimen occurs which causes deterioration of the image quality. Atkinson and McMillan (1974) described the use of cathodoluminescence to investigate a glass-ceramic microstructure. This technique depends on the fact that some crystals fluoresce when bombarded with electrons. By this method it was possible to obtain satisfactory micrographs of a glass-ceramic without the need of prior etching.

For investigation of phase separation or nucleation processes the higher resolution of the transmission electron microscope (TEM) is utilised. With this, features having dimensions of a few nanometres can be distinguished. Using thin specimens and utilising careful techniques, even higher resolutions can be achieved with the scanning transmission electron microscope (STEM); features only a few angstrom units in size are resolvable with this equipment. Identification of individual crystals by electron diffraction analysis or derivation of the chemical constitution of structural features by energy dispersive analysis of X-rays (EDAX) are further techniques that can be applied.

The use of direct transmission electron microscopy, which has largely superseded the older carbon replica technique, requires the preparation of a very thin section of the material, (of the order of 100 nm). A simple, but often effective, technique is to prepare the material in a finely powdered form and to "scan" this in the microscope until a wedge-shaped particle which is thin enough to transmit the electron beam is found. Generally, it is more useful, however, to prepare thin foils of the glass-ceramic under controlled conditions. The use of films of glass prepared by blowing and subsequently heat-treated to crystallise them is subject to a number of uncertainties and possible erroneous interpretation. Thin foils prepared by etching in dilute hydrofluoric acid can be of value especially for studies of glass-in-glass phase separation but if a high volume fraction of crystal phases is present the large difference in etching rates between crystalline and residual glass phases can result in disintegration of the specimen. The most satisfactory technique is first to prepare a thin section by mechanical methods such as diamond sawing, followed by lapping to achieve a thickness of a few tens of microns. The final thinning is then achieved by argon ion bombardment. Although the latter technique is comparatively slow (removal of material being at the rate of 1 micron per hour) it has the advantage that relatively large-area thin foils can be prepared allowing thorough characterisation of the glass-ceramic microstructures. Other techniques for preparing thin sections for electron microscopy involve the preparation of thin slivers of material by mechanical methods. One such technique makes use of the ultramicrotome. These methods do not appear to have been widely adopted however.

In studying the development of glass-ceramic microstructures it would be an advantage to observe the crystal nucleation and growth processes directly

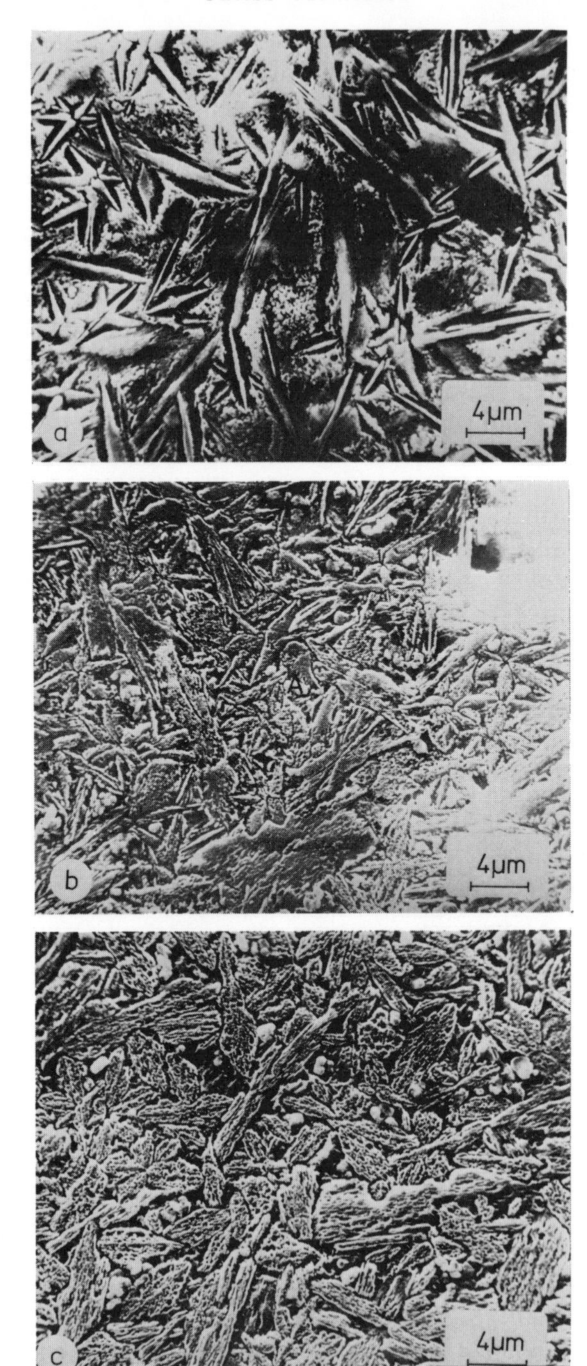
4μm
a
4μm
b
A
4μm
c

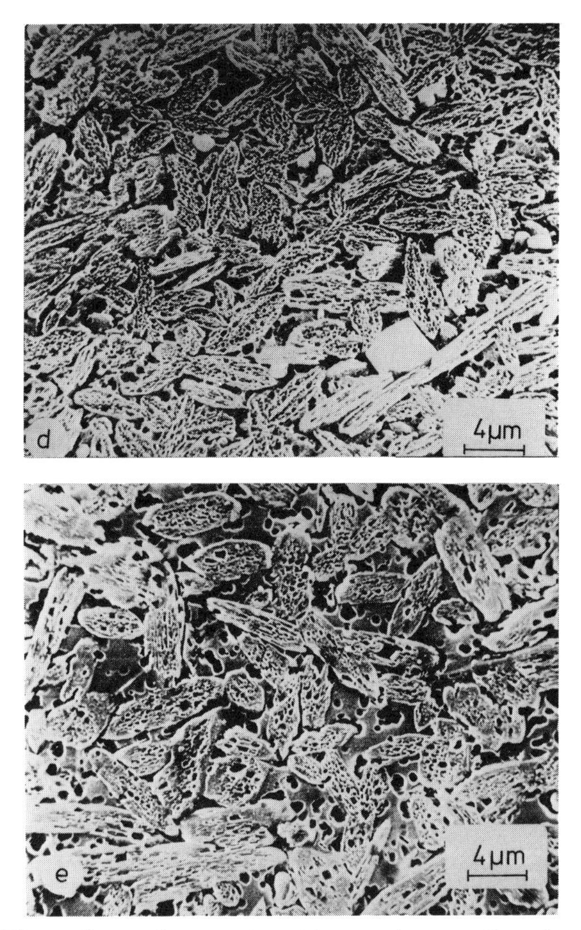

FIG. 39. Effects of upper heat-treatment temperatures on the microstructure of a glass-ceramic (a) 700°C, (b) 750°C, (c) 800°C, (d) 850°C, (f) 900°C.

and it is possible to make use of an electron microscope provided with a hot-stage. While the information gained in this way can be of value, the technique has serious limitations for the study of the development of glass-ceramic microstructures. Because the surface of volume ratio of the specimen is so large, both crystal nucleation and growth are dominated by surface effects and processes observed are unlikely to be the same as those taking place in a bulk specimen. A further difference is that the thin film specimen in the microscope is heat-treated *in vacuo* and there is evidence that both nucleation and growth kinetics can be different from those taking place in air. For these reasons it is more reliable to make use of specimens prepared from quenched samples taken at various stages of the heat-treatment process.

The value of electron microscopy for the study of microstructural development in glass-ceramics is illustrated by the SEM micrographs in Fig. 39 (Atkinson and McMillan, 1976). The glass-ceramic had a molecular percentage composition of SiO_2 61; Li_2O 30·5, K_2O 1·5, Al_2O_3 1·0, P_2O_5 1·0, B_2O_3 5·0. The micrographs show the effects on the glass-ceramic microstructure of 1-hour heat-treatments at temperatures in the range 700° to 900°C. It is apparent that the microstructural parameters are strongly dependent upon the heat-treatment temperature. This is clearly shown by Figs 40 and 41 which were derived by analysis of the electron micrographs. The volume fraction of the crystalline phases, maximises for a heat treatment temperature of 750°C. Also, the value of λ, the mean free path in the residual glass phase is a minimum for the glass-ceramic heat treated at 750°C. This parameter which is given by:

$$\lambda = d(1 - V_f)/V_f$$

where d is the mean crystal diameter, is important with regard to the mechanical strength of glass-ceramics as will be explained in chapter 5. It is also of interest to note that X-ray diffraction analysis showed the effects of heat-treatment on the nature of the crystal phases. At temperatures below 750°C, only one crystalline phase, lithium disilicate was formed having the needle-like morphology shown in Fig. 39a. At higher temperatures, a second phase identified as tridymite was formed.

c. Differential thermal analysis. Chemical reactions or structural changes within a crystalline or glassy substance are accompanied by the evolution or absorption of energy in the form of heat. When a substance crystallises, for example, an exothermic effect occurs since the free energy of the regular crystal lattice is less than that of the disordered liquid state. Conversely, the melting of a crystal gives rise to an endothermic effect. Chemical reactions between two substances may also give rise to endothermic or exothermic effects. Differential thermal analysis or D.T.A. is a technique which enables reactions or phase transformations to be studied for substances at high temperatures. In this method the material under test, in the form of a finely divided powder, is placed in a small capsule, often of platinum or other suitable refractory metal. Adjacent to the test capsule is a second capsule containing an inert power such as aluminium oxide which does not exhibit endothermic or exothermic effects. Thermocouples are embedded in the test substance and in the alumina powder and are connected so that their e.m.f.s are opposed; the net e.m.f. therefore represents the temperature difference between the sample powder and the inert alumina powder. The two capsules are heated together at a constant rate and the differential temperature is

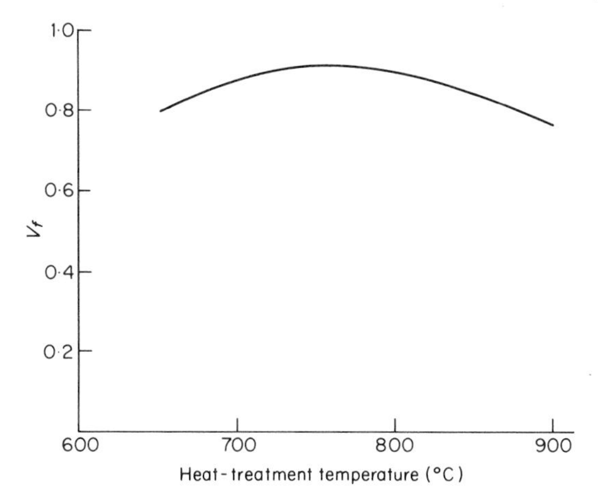

FIG. 40. Effect of heat-treatment temperature on volume fraction, V_f, of crystalline phase in a glass-ceramic.

plotted as the ordinate against the reference sample temperature as abscissa. Exothermic effects are indicated as peaks on the curve obtained and endothermic effects as dips in the curve.

If a sample of a glass which devitrifies on heating is treated in this way at least one exothermic peak will be observed corresponding with the separation

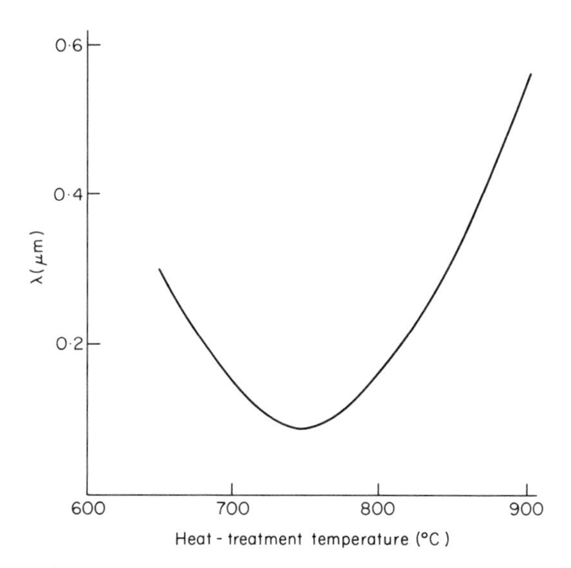

FIG. 41. Effect of heat-treatment temperature on mean free path, λ, for a glass-ceramic.

of a crystal phase. D.T.A. therefore offers a useful method of investigating the crystallisation of glasses and of determining the temperatures at which different crystals are formed. The D.T.A. curve obtained during the crystallisation of a glass usually shows a number of features and these can be seen in Fig. 42 which shows a typical curve.

As the temperature is increased, a dip is observed in the D.T.A. curve owing to a slight absorption of heat which occurs when the annealing point of the glass is reached. With further increase of temperature, one or more quite sharp exothermic peaks are observed, corresponding with the appearance of various crystal phases. At a higher temperature still, a marked endothermic effect is noticed and this is due to the first melting of the crystalline phases. Thus the D.T.A. curve can yield a great deal of very useful information which is of assistance in devising heat-treatment schedules for glass-ceramics, since it not only indicates the temperature ranges in which crystallisation occurs but it also indicates the maximum temperature to which the glass-ceramic could be heated without encountering deformation due to melting of crystal phases. Having determined the D.T.A. curve the exothermic peaks can be assigned to the crystallisation of various phases. For this, glass specimens are heated in turn to maximum temperatures corresponding to the exothermic peaks and are then subjected to X-ray diffraction analysis.

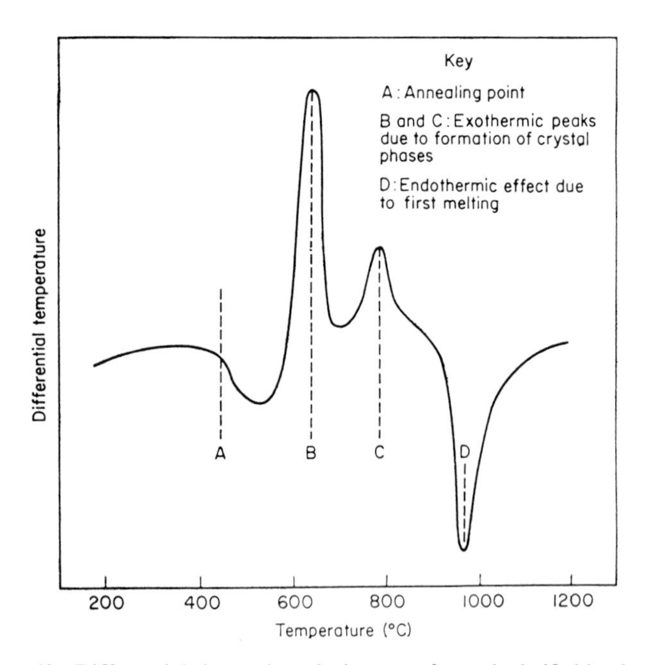

FIG. 42. Differential thermal analysis curve for a devitrifiable glass.

Stookey (1962) has described the use of D.T.A. to arrive at an optimum heat-treatment for special types of glass-ceramics. For these materials the aim was to produce ferroelectric compounds, such as barium titanate, as the predominant crystalline phases with the object of achieving high dielectric constants. The D.T.A. curves for these substances show a first dip in the range 500°C to 700°C, corresponding with the annealing ranges of the glasses. At about 50°C to 150°C above the annealing point a pronounced exothermic peak occurs, corresponding with the crystallisation of the ferroelectric compound as the primary crystalline phase. The separation of other crystalline phases is also indicated by exothermic peaks at somewhat higher temperatures. At a higher temperature still (usually in the range 1000–1300°C, depending upon the composition of the glass) the endothermic dip due to first melting of the crystalline phases is observed. It was shown that the optimum heat-treatment temperature for the development of ferroelectric crystals was midway between the exothermic peak, due to the ferroelectric phase, and a temperature about 50°C below the endothermic dip, due to first melting.

Thakur and Thiagarajan (1966) have described D.T.A. techniques from which information concerning the efficiency of nucleating agents can be assessed and whereby the complex activation energy of crystallisation E, made up by contributions from the activation energies of nucleation (E_n) and crystal growth (E_g) can be derived.

They derived an expression relating to the complex activation energy to the temperature of the exothermic D.T.A. peak, T_m as follows:

$$\log_e B/T_m^2 = \log_e (AR/E) - (E/R)(1/T_m)$$

where A is a frequency factor and R is the gas constant and B is the heating rate. From a plot of log B/T_m^2 versus $1/T_m$ the value of the complex activation energy could be derived. Using this method, the changes of complex activation energy for lithium disilicate glass as a function of platinum content between zero and 16×10^{-5} mole per 100 gm glass were determined. This indicated that the most effective nucleation (lowest complex activation energy) occurred for a platinum concentration of $\sim 8 \times 10^{-5}$ mole per 100 gm glass.

The same investigators also proposed another method for estimating the complex activation energy. For glasses in which surface nucleation predominates, the position of the exothermic crystallisation peak is strongly dependent on the surface area of the powder used. More finely divided powder causes a reduction of the exothermic peak temperature in such a case. If efficient bulk nucleation occurs, however, the effects of particle size upon the position of the exothermic peak will be small. Determination of the peak temperatures for standard "fine" and "coarse" powders therefore enables the efficiency of bulk nucleation to be estimated. Thakur and Thiagarajan argued that the difference in peak temperatures ΔT was related to the complex activation energy of crystallisation, E, an increase of ΔT indicating an increase of E. Their results for the platinum-nucleated lithium disilicate glass gave support to this idea.

Using the technique they were able to compare the efficiencies of a number of nucleating agents.

Briggs and Carruthers (1976) developed a D.T.A. technique for investigating the crystal growth kinetics of glasses. The compositions investigated were of the $CaO–MgO–Al_2O_3–SiO_2$ type and the results of D.T.A. studies were compared with those from hot stage microscopy. Using the latter technique they showed that the crystal growth rate V could be estimated from:

$$V = \exp(s) \exp(-q/T)$$

where s is a constant, q is the activation energy and T is the absolute temperature.

Assuming that ΔT the temperature difference arising at any point in time during the D.T.A. run was proportional to $\delta\alpha$ the incremental volume fraction of crystalline material produced, they were able to derive "theoretical" exotherms for a given glass. Comparison of these with the experimental exotherm allowed "best fit" values for s and q to be derived and hence allowed an estimate of crystal growth rates to be made from the D.T.A. data. It was suggested that this method was considerably more convenient and less time-consuming than the more conventional method of hot-stage microscopy.

d. Light scattering. If a beam of light enters a medium with a liquid-type structure, such as glass, scattering of light will occur in directions away from that of the incident beam due to point-to-point variations in the refractive index of the medium. Scattering centres can arise due to thermal fluctuations of density, but such centres are small in relation to the wavelength of light. If particles are present in the medium with somewhat larger diameters, however, the intensity of the scattered radiation will be very much increased. The early stages of nucleation and crystallisation of a glass establish conditions under which light scattering will occur, since small discrete particles differing in refractive index from the bulk of the glass are precipitated and the intensity of the scattered light will be related to the numbers and sizes of the particles. This means that quite a simple experiment, in which the intensity of scattered light is measured in a direction at 90°C to the incident beam, can detect the presence of scattering particles. Also, by comparative measurements for glass heat-treated under various conditions, an indication of the concentration of the particles can be obtained.

With a suitably refined apparatus much more precise information can be obtained concerning the size of the scattering particles and in detecting anisotropy of the particles such as would occur if the particles were crystalline. A light-scattering photometer suitable for experiments of this nature has been fully described by Brice *et al.* (1950). Very briefly, the equipment consists of a special cell in which a polished glass block specimen is immersed in a liquid of similar refractive index. A parallel beam of monochromatic light derived from a mercury vapour lamp provided with suitable filters enters the specimen and the intensity of scattered light at various emergent angles is measured by means of a photomultiplier tube. An analyser, usually a sheet of Polaroid, is interposed between the specimen and the photomultiplier so that the electric vector can be made vertical or horizontal for depolarisation measurements.

In experiments of this nature there are three important parameters which are determined; these are Rayleigh's ratio, dissymmetry and depolarisation. Rayleigh's ratio is given by:

$$R_u(90) = Ir^2/I_0V$$

where I is the scattered intensity in a direction at 90° to the incident beam; I_0 is the intensity of the incident unpolarised beam; r is the distance from the point of observation to the scatterer; and V is the scattering volume.

Dissymmetry is the ratio of the intensity of light scattered at 45°C to that scattered at 135°C.

Depolarisation is the ratio of light intensities of vertical and horizontal polarisation scattered at 90°.

The use of light-scattering experiments by R. D. Maurer (1962) to study crystal nucleation in glasses has already been mentioned in chapter 3. This work provides an excellent example of the value of this technique and the experimental results will therefore be described in some detail.

Maurer investigated the first stages of crystallisation of a glass of the approximate weight percentage composition: SiO_2: 49; MgO: 10; Al_2O_3: 30; TiO_2: 9; Na_20: 1·3; As_2O_5: 0·7. For comparison purposes a glass containing no titania but with the same relative proportions of the other constituents, and described as the base glass, was also prepared. Polished block specimens of the glass were heat-treated at temperatures between 725°C and 770°C for various lengths of time with intervening cooling to room temperature for scattering determinations.

Two important light-scattering parameters in this investigation were the dissymmetry of scattering, which is governed by the spatial variation of refractive index, and the depolarisation, which indicates the anisotropy of the refractive index.

For the lowest temperature heat-treatment, the depolarisation increased by a factor of six without any change in dissymmetry and with only a 30 per cent change in total scattering. Thus the spatial variation of refractive index remained constant while the scattering centres became more anisotropic. The interpretation of this result was that crystallisation of an emulsion-like phase was taking place. Prolonged heat-treatment at 770°C enabled a fairly clear X-ray diffraction pattern to be obtained, from which it was deduced that the crystalline phase formed was magnesium titantate.

Maurer showed that an average particle size could be deduced from measurements of the Rayleigh's ratio and the volumetric concentration of crystals obtained from X-ray diffraction measurements. Assuming that the number of crystals remains constant throughout the growth process, the crystal diameter varies as the sixth root of the Rayleigh's ratio. For the glass under investigation, it was deduced that the smallest particles detected were 25 Å in diameter and that the average particle size increased from 70 to 94 Å for heat-treatment at 725°C to 141 to 175 Å for heat-treatment at 770°C. Similar values were obtained independently from X-ray line broadening measurements.

The presence of an emulsion-like phase in the glass before heat-treatment was confirmed by the light-scattering measurements, since the scattering was higher by a factor of three than would be expected for a normal glass. Annealing of the glass did not significantly affect the scattering and it was concluded that the inhomogeneities responsible for scattering were due to two-phase separation of the melt during the early stages of cooling.

The importance of titania in promoting phase separation and crystallisation was demonstrated by comparing the behaviour of the titania-containing glass and that of the base glass. When the glasses were heated for 25 hours at temperatures at which their viscosities were $1{\cdot}5 \times 10^{12}$ poises (770°C for the titania-containing glass; 812°C for the base glass) the light-scattering changed by a factor of 10^5 for the titania glass whereas no detectable change occurred for the titania-free base glass.

e. The use of physical property measurements. As the glass structure is transformed into that of the polycrystalline ceramic it is clear that significant changes of physical properties will take place. Measurement of suitable properties therefore, either during the heat-treatment cycle itself or on specimens quenched from various points in the cycle can provide a sensitive and convenient way of monitoring the structural changes.

The dimensional change taking place during the heat-treatment cycle, although small, can be measured by a suitable dilatometer or other means. McMillan *et al.* (1966a) observed the changes of length of fibres of a glass-ceramic during the heat-treatment process. The results showed that a marked contraction of length amounting to about 1·67 per cent occurred during crystallisation.

It has also become apparent that measurement of electrical properties provides a very sensitive method of detecting structural change and can

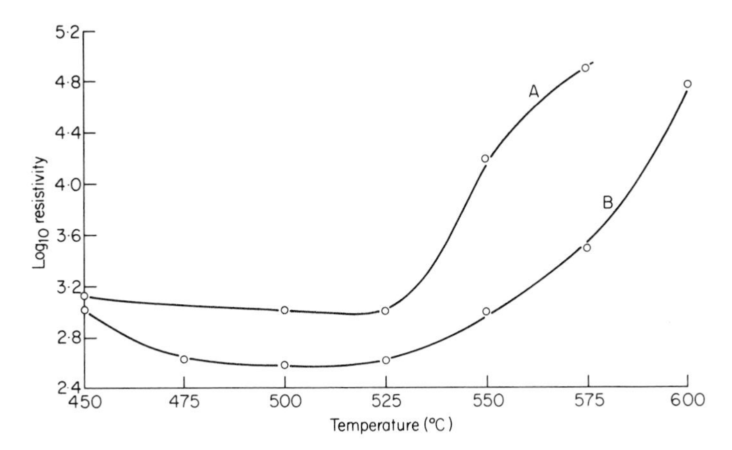

FIG. 43. Resistivity values for glass-ceramics at 350°C after heat-treatment at the indicated temperatures. Molecular percentage compositions:
A SiO_2 70; Li_2O 30 B SiO_2 69: Li_2O 30; P_2O_5 1.

indicate the onset of crystallisation before it is possible to detect the formation of crystal phases by X-ray diffraction techniques.

In glasses in which the mobile cations responsible for electrical conduction (usually alkali metal ions) are incorporated in a crystal phase, it is to be expected that the conductivity will be reduced when the glass crystallises and similarly for the dielectric loss tangent. This is because the energy barriers that the ions must surmount in order to be transported or undergo oscillatory motion are less for the disordered glass structure than for the regular crystal lattice.

Figure 43 shows the results of measurements of electrical resistivity made on specimens of two lithia-silica compositions heat treated at temperatures in the range 450 to 600°C. (McMillan, 1971). Measurements of resistivity were made at 350°C. Although the results indicate the occurrence of crystallisation at temperatures of 550°C and upwards, crystallisation was not detected by X-ray diffraction until the glasses were heat-treated at 600°C, when lithium disilicate was detected. The results suggest that the crystallisation takes place more slowly for the P_2O_5-containing glass, a fact that could have been demonstrated also by quantitative X-ray diffraction analysis but this would be much more time-consuming.

Measurement of changes of conductivity during isothermal treatment (Phillips and McMillan, 1965) give valuable information as shown in Fig. 44. For both glasses, the electrical resistivity increases with time but the slower rate of increase, especially for the lower heat-treatment temperature of 540°C, suggests that the P_2O_5-containing glass crystallises more slowly.

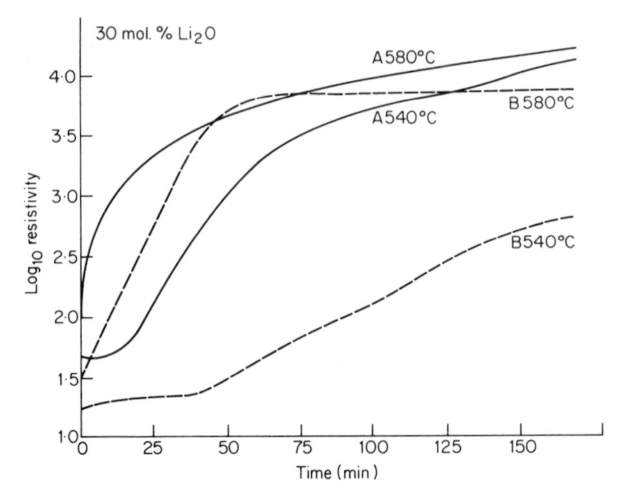

FIG. 44. Resistivity as a function of duration of heat-treatment for two glasses (compositions as given in Fig. 43).

Measurements of the dielectric loss tangent of lithia-silica glasses by Harper *et al.* (1970) showed a very marked reduction of this parameter for specimens heat treated at 600°C and above. This temperature corresponds with the development of lithium disilicate crystals (Fig. 45). Further measurements under isothermal conditions enabled the volume fraction V_f of the crystalline lithium disilicate phase to be estimated assuming that the loss tangent of the

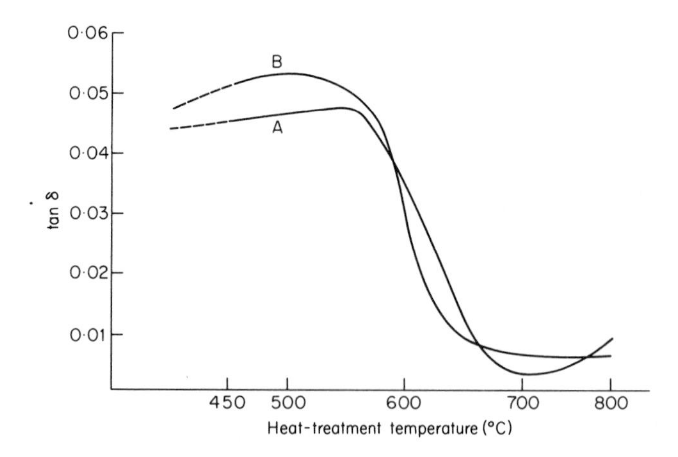

FIG. 45. Loss tangent at 1 MHz as a function of heat-treatment temperature for glasses having the molecular percentage compositions:
A SiO_2 70; Li_2O 30 B SiO_2 69; Li_2O 30; P_2O_5 1.

glass-ceramic after time t, tan δ_t was an additive function of the loss tangents of the crystalline phase, tan δ_{cr}, and the residual glass phase, tan δ_g. The expression

$$V_f = \tan\delta_g - \tan\delta_t / \tan\delta_g - \tan\delta_{cr}$$

was derived to calculate V_f. The results obtained in the isothermal treatment of four glasses at 600°C are given in Fig. 46. It is of interest to note that while two of the glasses showed an induction period before crystallisation began, the other two glasses commenced to crystallise immediately at 600°C. The fact that the V_f versus time curves are linear suggests fibrillar growth since 2-dimensional (plates) or 3-dimensional growth would have given a quadratic or cubic dependence respectively. It is also evident that the rates of crystallisation of glasses 1 and 2 were significantly lower than those for the other glasses. This may be connected with the fact that the former were two-phase glasses while the latter did not undergo glass-in-glass phase separation.

An interesting method using an electrical measurement to investigate glass crystallisation was developed by Baak (1967). He showed that if a specimen of a crystallisable glass is heat-treated in contact with a specimen of the same composition in its fully crystallised state, an electromotive force is generated between the two which is dependent on the crystallisation process taking place. Baak was able to correlate the measured e.m.f. with the percentage crystalline content as shown in Fig. 47 and this method might therefore be useful for monitoring the progress of the crystallisation process. It appears, however, that the method is more sensitive for the lower volume fractions of crystalline content (ie below 50 or 60 per cent). Baak suggested that the origin of the effect was the transference of oxygen ions from the glass specimen to the crystallised specimen.

Changes of viscosity as a result of crystallisation are also valuable in providing indications of the changes of structure taking place. An ingenious application of viscosity measurements to investigate the crystallisation of Na_2O–Nb_2O_3–SiO_2 glasses was made by Layton and Herczog (1967). The viscosity of the glasses was initially found to decrease but above a certain temperature depending on the glass composition it increased owing to crystallisation. Curves 1 and 2 in Fig. 48 illustrate this effect. The authors proposed that increase of the silica content of the residual glass, resulting from the precipitation of sodium niobate crystals, was largely responsible for the increase of viscosity. They assumed that the viscosity was little affected by the dispersed crystalline particles but it should be pointed out that this assumption may be less reliable if the volume fraction of crystals rises above some relatively low value. By drawing viscosity temperature curves for glasses of higher silica contents, which did not crystallise until much higher temperatures were attained (eg curves 3 and 4 in Fig. 48) Layton and Herczog

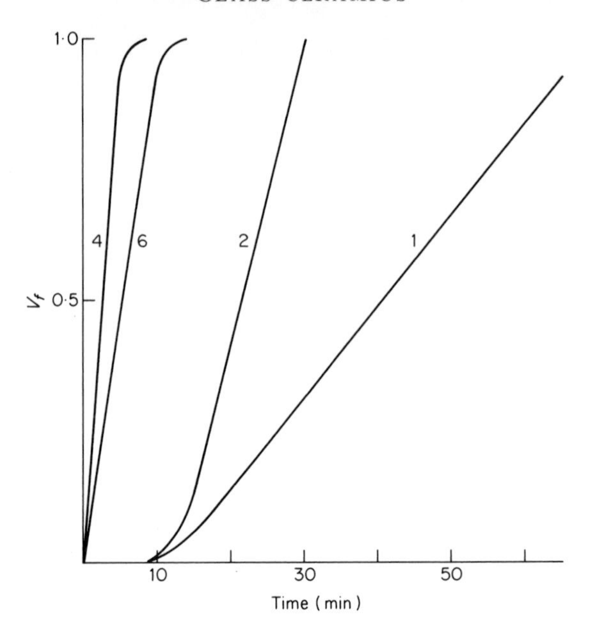

FIG. 46. Volume fractions of crystal phase as a function of duration of heat-treatment duration at 600°C for Li_2O–SiO_2–P_2O_5 glass-ceramics (compositions are given in Table VI).

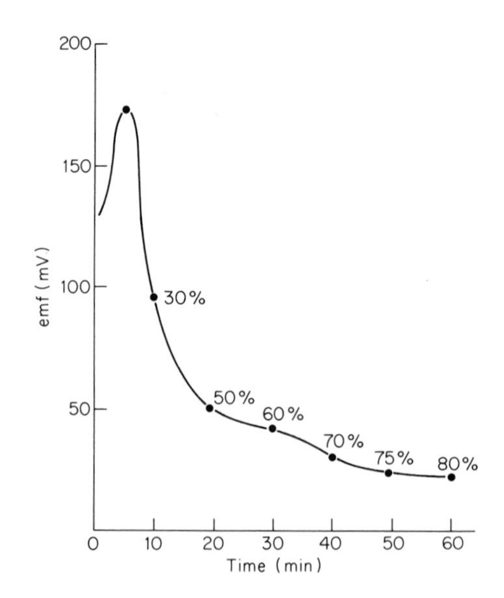

FIG. 47. E.m.f. as a function of heat treatment for a crystallising glass (after Baak, 1967).

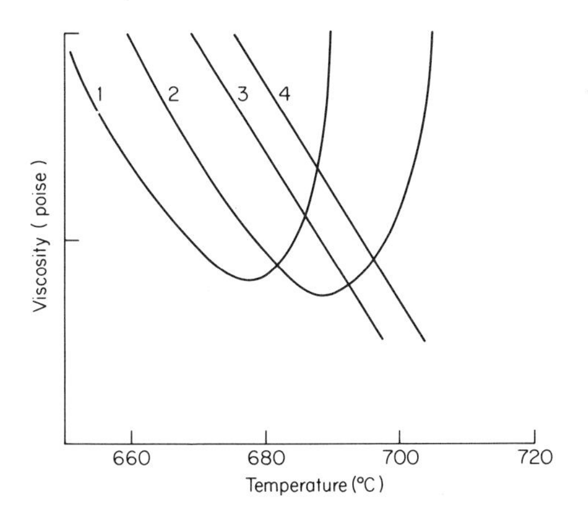

FIG. 48. Viscosity-temperature curves after Layton and Herczog (1967) for Na_2O–Nb_2O_3–SiO_2 glass-ceramics containing:
A 19·4 wt per cent SiO_2 C 23·4 wt per cent SiO_2
B 21·0 wt per cent SiO_2 D 26·7 wt per cent SiO_2.
(Reproduced by courtesy of the American Ceramic Society.)

were able from the intercepts to deduce the silica contents of the partially crystallised glasses after heating to various temperatures. From this information, it was possible to calculate the crystalline contents of the glass-ceramics and to demonstrate reasonable agreement between these and values derived from density measurements.

McMillan *et al.* (1966b) used measurements of viscosity and electrical resistivity to study the crystallisation process in a glass essentially of the Li_2O–ZnO–SiO_2 type containing P_2O_5 as a nucleating agent. The results were interpreted in terms of the metastable formation of lithium metasilicate and the conversion of this to lithium disilicate when the temperature rose above about 560°C.

Another physical technique that has potentialities for investigating the crystallisation of glasses is infrared spectroscopy. Absorption spectra of thin films may be used, but a more convenient technique is that of attentuated total reflectance (ATR) spectroscopy by which the structure of a thin surface layer of bulk specimens can be probed. The type of information that can be derived from the two methods can be assessed from Figs. 49 and 50. The former shows absorption spectra for a quenched specimen of a phosphate-nucleated Li_2O–SiO_2 glass together with that for a specimen heat treated at 550°C for 0·5 hr. Marked changes resulting from crystallisation are apparent, not only in the sharpening of absorption bands but also in the appearance of new bands. The intensity of the absorption band at 16 μm correlates with the duration of

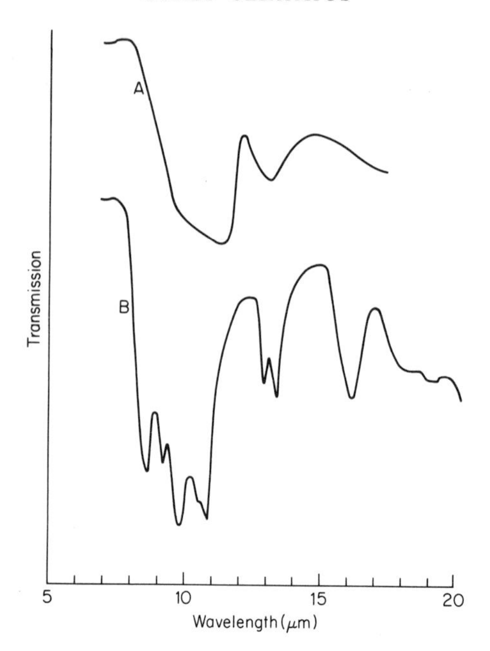

FIG. 49. Infrared transmission curves for glass SiO_2 69; Li_2O 30; P_2O_5 1. Curve A, quenched glass; Curve B heat treated at 550°C for 30 minutes (for clarity the curves are displaced vertically).

heat treatment at 550°C and therefore gives a measure of the progress of crystallisation.

ATR spectra for the same glass are given in Fig. 49. The unpolished specimen of the crystallised glass (curve B) gave a spectrum showing essentially the same features as the absorption spectrum. Polishing of the specimen (curve C) however removed the "crystalline" features and resulted in a spectrum very similar to that for the uncrystallised glass (curve A). The results clearly showed that surface crystallisation was taking place under the conditions of heat-treatment employed.

3. *Crystal Types and the Conditions under which they are Formed*

Although phase equilibrium diagrams have been published (Levin *et al.*, 1956, 1959 and 1964) for some of the oxide systems which are important for the production of glass-ceramics, these diagrams serve as a general guide only and cannot be used to predict accurately the nature or proportions of the phases which will be present in a glass-ceramic. In the high silica regions of certain systems, for example, tridymite would be expected to appear as a major phase

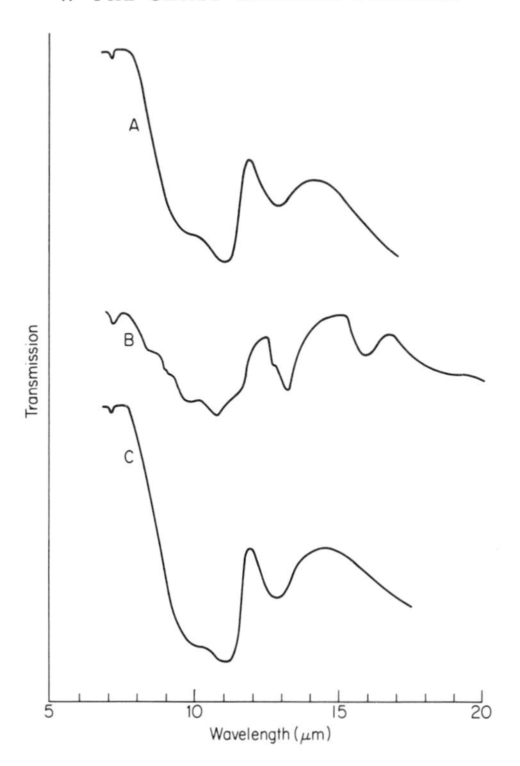

FIG. 50. A.T.R. curves for glass SiO_2 69; Li_2O 30: P_2O_5 1.
Curve A, untreated glass; curve B, heat treated at 550°C for 30 min; Curve C, heat treated as for "B" but subsequently ground and polished (for clarity the curves are displaced vertically).

according to the published phase diagrams. It is fairly unusual to obtain this form of crystalline silica in a glass-ceramic, however, and the free silica more often occurs as cristobalite or quartz. The usual phase diagrams have been determined for conditions of temperature and time which are not applicable in the glass-ceramic process, so that it is perhaps not surprising that they cannot be used as an accurate guide to the phases which will be present in a particular glass-ceramic composition. It would be extremely difficult to produce a comprehensive phase diagram for a particular glass-ceramic system since the crystalline phases present can be different even for material of the same chemical composition if the heat-treatment schedule used to accomplish crystallisation is varied. The time the glass spends at the lower temperatures before it is raised to the maximum crystallisation temperature can have a marked effect upon the nature of the crystal phases which are ultimately developed. In addition, minor changes of composition often exert an apparently disproportionate effect upon the crystal phases present in the glass-ceramic. Some principles have been established however which provide a useful guide concerning the conditions under which different crystal phases are formed and the important glass-ceramic systems will be discussed in the light of these.

The Li_2O–SiO_2 system may be regarded as the prototype of several important glass-ceramic systems and the crystallisation of glasses of this type is therefore of interest. As mentioned earlier, measurements of electrical resistivities of such glasses indicated that crystallisation commences at a temperature in the region of 550°C although X-ray diffraction measurements only indicated the occurrence of crystallisation at temperatures of 600°C and upwards. Later work by James and McMillan (1971) revealed that heat treatment of a P_2O_3-containing glass at 600°C resulted in crystallisation as revealed by transmission electron microscopy. Selected area electron diffraction (S.E.D.) showed that the crystals were lithium disilicate. The development of this phase at an even lower temperature was demonstrated by Hing and McMillan (1973a) using high resolution transmission electron microscopy in combination with selected area diffraction. It was shown that while there was no evidence of crystallisation in the binary 30 Li_2O 70 SiO_2 glass after heat-treatment at 500°C, glass-in-glass phase separation had occurred. The replacement of 1 mol. per cent SiO_2 by P_2O_5 resulted in the occurrence of crystallisation at 480°C. After heat-treatment for 1 hour at 520°C, this glass contained crystallites of sizes ranging from 5 to 20 nm and S.E.D. showed that these were lithium disilicate. The crystallites could not be detected by X-ray diffraction at this stage. Inclusion of 1 mol. per cent ZnO in the glass resulted in coarse and extensive glass-in-glass phase separation and also led to cracking of the material after heat-treatment at 550°C. The further inclusion of 1 mol. per cent K_2O was made with the aim of lowering the viscosity of the residual glass phase and thereby avoiding the cracking. The resultant material was shown to contain lithium disilicate in a microcrystalline form. This work illustrates the effects of minor constituents in influencing the crystallisation process in addition to the recognised effects of nucleating agents.

The effects of the latter on the crystallisation of Li_2O–SiO_2 glasses containing between 25 and 35 mol. per cent Li_2O have been investigated. The nucleation catalysts employed were P_2O_5, TiO_2 and metallic platinum. In the absence of a catalyst, the formation of lithium silicate could be detected after heat treating the glasses at temperatures ranging from 557°C to 569°C. This temperature range was apparently unaffected by any of the nucleation catalysts. A second crystalline phase, quartz, developed at a temperature between 615°C and 690°C. While the inclusion of TiO_2 did not affect the temperature at which quartz developed, the incorporation of P_2O_3 caused a reduction of the temperature to between 585°C and 595°C and platinum exerted a similar but less marked effect.

There seems to be clear evidence that while P_2O_5, and to a lesser extent platinum, catalyse the crystallisation of quartz, TiO_2 is not effective in this respect. The superiority of P_2O_5 in promoting nucleation is underlined by the finding that the compositions in which it was included gave microcrystalline glass-ceramics of high strengths while those containing TiO_2 or no nucleating agent were weak and coarsely crystalline. The materials containing platinum were also microcrystalline but were less strong than those containing P_2O_5.

Another important binary glass-ceramic system comprises the Al_2O_3–SiO_2 glasses studied by MacDowell and Beall (1969). The compositions had Al_2O_5 contents ranging from 5 to 60 mol. per cent and were heat-treated at temperatures between 900°C and 1600°C.

The glasses became less stable as the Al_2O_3 content increased and for contents higher than about 30 mol. per cent very fast cooling was necessary to produce glassy materials. All of the glasses, with the exception of the 60 mol. per cent Al_2O_3 composition, showed metastable immiscibility on cooling and the two-phase structure coarsened during subsequent heat treatment. In some cases the alumina-rich phase crystallised to mullite on cooling. Inclusion of small amounts of modifying oxides such as Na_2O, CaO or BaO tended to suppress two-phase separation during cooling and melts containing up to 50 mol. per cent Al_2O_3 could be quenched to homogeneous glasses. The inclusion of network forming cations (eg B^{3+}, Ti^{4+}, Ge^{4+}) increased the scale of phase separation and enhanced the tendency toward devitrification of compositions containing less than 23 mol. per cent Al_2O_3. Heat treatment of the compositions containing less than about 20 per cent Al_2O_3 caused the development first of mullite in a finely divided form. For example, the glass containing 15 mol. per cent Al_2O_3 when heat treated at 950°C for 10 hours developed the X-ray diffraction pattern for this phase although the material appeared glassy and transparent. Gradual increases in the coefficient of thermal expansion from $13{\cdot}4 \times 10^{-7}$°C^{-1} for the uncrystallised glass to $18{\cdot}9 \times 10^{-7}$°C^{-1} for the material heat treated at 1150°C and increases of density suggested a gradual increase in the amount of mullite phase as the heat-treatment temperature was increased. When the heat-treatment temperature was raised to 1200°C, a dramatic increase of the thermal expansion coefficient to $139{\cdot}3 \times 10^{-7}$ occurred and this was due to the precipitation of cristobalite in addition to the major phase, mullite. Even after heat-treatment at this temperature the glass-ceramic remained transparent indicating the very fine scale of the microstructure. Heat-treatment at 1500°C led to the presence of cristobalite as a major phase having a spherulitic morphology and resulting in a high thermal expansion coefficient of

$260{\cdot}6 \times 10^{-7}$. This glass-ceramic was white and opaque although still having a very fine-grained microstructure.

The crystallisation behaviour of the glasses was modified by the inclusion of further oxides to give ternary compositions. For example, a glass containing 30 mol. per cent Al_2O_3 in which 5 mol. per cent BaO was included, showed essentially complete crystallisation of mullite after heat-treatment at 950°C and the formation of cristobalite at heat-treatment temperatures of 1200°C upwards was suppressed. Small additions of other alkaline earth and alkali metal oxides had a similar effect.

This investigation provides a valuable example of the influence of two-phase separation upon crystallisation of glasses and also of the marked changes of behaviour that can result from comparatively minor additions of suitable oxides.

The Li_2O–Al_2O_3–SiO_2 system is important since it is from the high alumina region of this that low thermal expansion glass-ceramics are derived. Also compositions having relatively low Al_2O_3 contents form the basis for photosensitively nucleated glass-ceramics. Eppler (1963) undertook a useful survey of the crystallisation of glasses from this system. He found that only glasses containing more than 20 weight per cent Li_2O showed crystallisation when heat treated at 600°C. Lithium disilicate was formed when the silica content exceeded about 75 per cent but for lower silica contents the major phase was lithium metasilicate. For compositions having silica contents less than 70 to 75 per cent a beta-eucryptite/quartz solid solution was also formed. For heat treatments at 700°C and upwards lithium metasilicate occurred as a major phase for practically all of the compositions studied. At these temperatures a beta-spodumene/quartz solid solution also developed for most compositions with the exception of the high silica, low alumina compositions. This phase, it was considered, formed by recrystallisation of the beta-eucryptite/quartz solid solution rather than by direct crystallisation from the glass.

For glasses of the Li_2O–Al_2O_3–SiO_2 type containing P_2O_5 as a nucleation catalyst it is found that in compositions containing less than about 10 per cent Al_2O_3 lithium silicates appear as major phases; the disilicate forms in glasses containing more than 70–75 per cent SiO_2 and the metasilicate in glasses of lower SiO_2 content. A crystalline form of silica often appears as a secondary phase in the silica-rich compositions and this may be quartz, cristobalite or a mixture of the two depending on heat-treatment and the presence of certain minor constituents, notably the oxides of alkali metals. Potassium oxide, for example, has a marked effect in favouring the formation of quartz even when present in concentrations of only 1 to 2 per cent and there is evidence that the oxides of higher members of the alkali metal group are still more effective. The avoidance of cristobalite formation is desirable in practical glass-ceramics since the α-β cristobalite inversion occurring in the region of 200°C is accompanied by a large volume change which can result in cracking of the glass-ceramic.

As pointed out by Beall *et al.* (1967) glasses of the aluminosilicate type from which β-quartz solid solutions may be precipitated as major crystalline phases provide a good example of a fairly common phenomenon namely, that a series of crystalline states may be traversed before the equilibrium constitution is achieved.

The basic principles governing the formation of aluminosilicate crystals having structures close to those of crystalline forms of silica were first formulated by Buerger (1954). The essential feature is that AlO_4 tetrahedra replace SiO_4 groups in the crystal lattice and in order to maintain overall electroneutrality, monovalent or divalent cations are accommodated in the interstitial sites. Quartz, cristobalite, tridymite and keatite which are crystalline modifications of silica all have aluminosilicate equivalents. These include carnegieite and nepheline ($NaAlSiO_4$) having the cristobalite and tridymite structures respectively. β-Spodumene has the keatite structure and β-eucryptite is a derivative of the β-quartz structure. In the case of β-eucryptite, the interstitial cation is lithium and half of the SiO_4 groups are replaced by AlO_4 groups. This crystal can form a whole series of solid solutions with silica which are isostructural with β-quartz.

β-Quartz solid solutions are also formed in the MgO–Al_sO_3–SiO_2 system with the divalent magnesium ions occupying the interstial sites. To a limited extent zinc may also participate in such structures both in the interstitial (octahedral) sites and in the tetrahedral sites thus exhibiting a dual role. Generally, however, the zinc ions only assume tetrahedral co-ordination if there is a deficiency of aluminium in the composition. Beryllium ions may also take up this structural position if aluminium is deficient. Conversely, if excess aluminium oxide is present in the composition, a proportion of the aluminium ions can enter the octahedral sites.

β-Quartz solid solutions are not stable at high temperatures and break down to yield other phases if heated sufficiently at sub-solidus temperatures. These breakdown products can include various spinels, sapphirine, cristobalite, mullite, willemite and lithium silicates. The nature of breakdown products has important consequences regarding the properties of glass-ceramics derived from β-quartz compositions. Two main groups of materials exist:

(*i*) Compositions in the MgO–ZnO–Al_2O_3–SiO_2 system having Li_2O contents less than 2·5 wt per cent. The breakdown products in this case comprise spinels plus siliceous β-quartz. On cooling the latter can invert to α-quartz. Thus glass-ceramics of this type have high coefficients of thermal expansion.

(*ii*) Compositions in the Li_2O–MgO–ZnO–Al_2O_3–SiO_2 system having

Li_2O contents greater than 2·5 weight per cent. In this case the β-quartz type crystals transform mainly into β-spodumene and thus these glass-ceramics have low thermal expansion coefficients.

Doherty *et al.* (1967) described an investigation of the crystallisation of a glass of the weight percentage composition:

SiO_2	69·5
Al_2O_3	17·8
MgO	2·8
Li_2O	2·5
TiO_2	4·75
ZrO_2	0·25
ZnO	1·0
Na_2O	0·4
As_2O_3	1·0

This study illustrated very clearly the principles discussed by Beall *et al.* (1967). Using X-ray diffraction it was shown that the first detectable crystalline phase was β-eucryptite which appeared at about 850°C. When the temperature was raised above 975°C the proportion of this phase diminished fairly rapidly and increasing amounts of β-spodumene appeared as the temperature increased to the maximum of 1100°C.

Transmission electron microscopy revealed an important feature namely that a large number of particles, about 5 nm in size and in a concentration of 1 to 2 volume per cent, were precipitated at about 825°C (ie just before the formation of β-eucryptite could be detected). The electron microscope also revealed microstructural changes as the heat-treatment temperature was progressively increased to 1100°C. From an initial size of 50 to 80 nm the β-eucryptite crystals formed at 875°C increased in size to $\sim$ 100 nm at 950°C. Heating to 1050°C (which exceeded the β-eucryptite $\rightarrow$ β-spodumene transformation temperature) resulted in crystals of average size 600 nm comprising 90 per cent volume fraction of the glass-ceramic. Increasing the heat temperature to 1100°C resulted in an almost totally crystalline material comprised largely of 1 μm β-spodumene crystals with some minor phases. One of these was rutile (TiO_2) present in a volume percentage of 2 to 3.

Careful analysis revealed that the small particles which preceded the β-eucryptite formation had a structure closely resembling the pyrochlore structure and in all probability comprised aluminium titanium pyrochlore $Al_2Ti_2O_7$. It seems probable that this phase, appearing in a finely dispersed form, may have acted as a heterogeneous nucleant for the crystallisation of β-eucryptite.

Another useful study of glass-ceramics of the Li_2O–Al_2O_3–SiO_2 type but having a higher Li_2O content was reported by Barry *et al.* (1969) who investigated crystallisation of glasses having weight percentage compositions in the range: Li_2O 17·5 to 19·5; Al_2O_3 13·4 to 15·0; SiO_2 55·9 to 65·7; TiO_2 0 to 11.

Differential thermal analysis showed that a marked exothermic peak occurred about 100–200°C above the transformation temperature (T_g) of the glass. This was attributed to the simultaneous crystallisation of β-eucryptite and lithium metasilicate. X-ray diffraction analysis indicated that the temperature range in which the former phase developed was 600 to 670°C and that for the latter was 650–680°C. In the glasses containing TiO_2, lithium titanate was also developed. Generally this required heat treatment at 700°C or higher although for the composition containing the maximum concentration of TiO_2, heat treatment at 600°C was sufficient to cause development of the titanate.

Glass-ceramics of the magnesia-alumina-silica type are of importance because they combine good electrical insulation characteristics with high mechanical strength and medium to low thermal expansion coefficients. These desirable properties arise largely as a result of the presence of alpha-cordierite ($2MgO\ 2Al_2O_3\ 5SiO_2$) as a principal phase in glass-ceramics of this type. This crystal appears upon heat-treatment of compositions throughout the whole glass-forming region. Other phases which may be present include cristobalite in the high silica compositions and clino-enstatite ($MgO\ SiO_2$) or forsterite ($2MgO\ SiO_2$) for compositions having high magnesium oxide contents.

Karkhanavala and Hummel (1953) discussed the crystalline modifications of cordierite. Three forms appear to exist: namely, the stable high temperature form (α-cordierite); a metastable low temperature form (μ-cordierite) and a stable low temperature form (β-cordierite). The latter could only be produced hydrothermally at temperatures below 830°C and its occurrence is therefore not important in relation to glass-ceramics. Both the μ and β forms were found to convert to the stable α-form when heat treated at sufficiently high temperatures. For μ-cordierite, conversion occurred relatively sluggishly at temperatures above 900°C and for β-cordierite at temperatures above 830°C. Karkhanavala and Hummel reported that μ-cordierite was not easily developed by the crystallisation of glasses and required many hours of heat treatment of finely powdered glass at temperatures in the region of 800–900°C. In contrast, the α-modification formed readily by the crystallisation of glass in the temperature range above 1050°C.

Later studies by a number of workers has indicated that these findings may not be fully applicable to more complex glass compositions. The presence of nucleating agents and of minor oxide constituents can significantly affect the crystallisation kinetics. As a result, μ-cordierite can often appear readily during the crystallisation of some compositions, although the probability of its formation is enhanced by the use of powdered rather than bulk glass, suggesting that surface nucleation may be an important factor. Also, the transformation from the μ to α forms at temperatures above 1050°C does not always appear to be sluggish and occurs reasonably rapidly for some glass compositions.

Some of the effects that occur are illustrated by the results of a study of a platinum-nucleated $MgO–Al_2O_3–SiO_2$ glass. The D.T.A. curve showed two exothermic peaks at temperatures of 900°C and 980°C. These peaks were shown by X-ray diffraction analysis to be due to the formation of μ-cordierite (900°C) and α-cordierite (980°C). Heat-treatment of bulk glass specimens to a maximum temperature of 900°C resulted in the development only of μ-cordierite. Heat-treatment at higher temperatures resulted in the formation only of α-cordierite owing to the conversion of the μ-form to the α-form.

Schreyer and Schairer (1961) investigated the polymorphism of cordierite and made an important contribution concerning the structure of μ-cordierite. They showed that it corresponds to a "stuffed" β-quartz (crystals of this type have already been discussed on p. 137). Studies have shown that this phase forms metastably as the first crystal phase to appear during the crystallisation of a wide range of compositions in the $MgO–Al_2O_3–SiO_2$ system. The crystallisation behaviour at low temperatures, which results in the formation of this metastable quartz-like crystal phase is clearly relevant to any consideration of the controlled crystallisation of cordierite-type glasses.

Gregory and Veasey (1971, 1972, 1973) provided a very comprehensive review and discussion of experimental studies of cordierite glass-ceramics. It was pointed out that the relative amounts of quartz-like and cordierite phases depended both on the silica content of the glass and on the heat-treatment temperature. An inverse relationship was seen to exist between the concentrations of the two forms clearly suggesting that the cordierite phase developed by rearrangement of the hexagonal structure of the metastable phase to give hexagonal cordierite. In their experimental studies, Gregory and Veasey used D.T.A. and infrared spectroscopy in conjunction with X-ray diffraction to study the effects of low temperature treatments on the subsequent crystallisation behaviour of cordierite glasses. It was shown, for example, that heat-treatment at 830°C for periods ranging from 1 to 24 hours had a profound effect upon the crystallisation behaviour of the glass at higher temperatures. The development of an exothermic peak at 935°C attributed to the formation of the metastable quartz-like phase, was enhanced by the prior treatment at 830°C.

These and similar studies underline the importance of careful control of the heat-treatment cycle for glass-ceramics based on the cordierite system if materials having reproducible properties are to be achieved.

In the $Li_2O–MgO–SiO_2$ system it is found that glass-ceramics having low

MgO contents bear similarities with regard to the crystal phases present and properties to the corresponding compositions derived from the Li_2O–Al_2O_3–SiO_2 system. In compositions having high silica contents, quartz is often present as a major phase although under certain conditions of heat-treatment cristobalite may also be formed; tridymite is rarely found in these materials. Depending upon the silica content, lithium disilicate or lithium metasilicate may also appear as crystal phases. For compositions in the high MgO regions of the glass-forming area, forsterite and clino-enstatite may be formed and for compositions having intermediate MgO contents lithium magnesium silicate (Li_2O MgO SiO_2) may appear. There are few data concerning the temperatures at which the crystal phases are developed in the Li_2O–MgO–SiO_2 glass-ceramics although it is known that lithium disilicate crystallises at a temperature in the region of 600°C and the crystalline forms of silica in the region of 700°C.

The Li_2O–ZnO–SiO_2 system is of great interest since in certain composition ranges hitherto unknown crystal types are developed. These phases are Li_2O ZnO SiO_2 which occurs in glass-ceramics having zinc oxide contents in the region of about 20 to 35 per cent ZnO and $4Li_2O$ 10ZnO $7SiO_2$ which occurs in glass-ceramics having ZnO contents higher than about 30 per cent. In compositions containing more than about 45 per cent ZnO willemite (2ZnO SiO_2) is sometimes formed. Lithium disilicate, together with quartz or cristobalite, appear as principal phases in compositions having ZnO contents less than about 10 per cent. Differential thermal analysis of a composition of this type has shown that lithium disilicate crystallises at a temperature in the region of 500°C to 570°C and the presence of phosphorus pentoxide in the glass does not greatly alter this temperature. Silica crystallises out at 720 to 740°C in P_2O_5-free compositions but at 670 to 700°C when this oxide is present. The glass-ceramics containing P_2O_5 are micro-crystalline and strong whereas those not containing this oxide are very weak and have a coarse microstructure. Thus P_2O_5 is an effective nucleating agent in these materials. Morell (1970) has studied the crystallisation of glass of this general type and shown that following the precipitation of lithium disilicate, the glass surrounding these crystals becomes enriched in zinc ions. Conditions then arise that are favourable for the nucleation and growth of a secondary phase containing zinc on the glass-crystal interface of the primary phase. The secondary phase is β-Li_2O, ZnO, SiO_2 and precipitation of this progressively depletes the glass phase of lithium oxide so that further crystallisation to produce zinc silicate and silica crystals becomes possible.

The work of Burnett and Douglas (1971) though on one of the less well known glass-ceramic systems illustrates very clearly the complexities that can occur in glass crystallisation. They investigated glasses of the Na_2O–BaO–SiO_2 system and showed that the first detectable crystal phase is a high temperature polymorph of barium disilicate which later converts to the stable low temperature form. This illustrates a feature that is not uncommon, namely, the formation first of a metastable phase followed by conversion to a stable phase. Another feature illustrated by their work is that for glasses of high silica content liquid unmixing takes place resulting in a silica-rich droplet phase; again, this phenomenon is often found in the development of glass-ceramics microstructures.

Heat treatment of a composition lying on the barium disilicate-sodium disilicate join caused barium disilicate to crystallise out and ultimately the composition of the residual glass phase is almost pure sodium disilicate which fails to crystallise; this glass is soluble in water. For some compositions, precipitation of barium disilicate causes the composition of the residual glass phase to move out of the region of immiscibility and the SiO_2-rich droplets redissolve causing the glass composition to diverge towards a higher silica content. This then results in the precipitation of a crystalline form of silica.

To form a glass-ceramic that is fully crystalline, the residual glass left after crystallising the primary crystalline phase should lie outside the glass-forming region of the system. In the Na_2O–BaO–SiO_2 system such a composition is 60 SiO_2 30 BaO 10 Na_2O. Unfortunately, crystallisation of the residual glass phase after precipitation of barium disilicate, although it results in a highly crystalline material, yields sodium silicate crystal phases which are readily attacked by atmospheric moisture.

A wide variety of crystal types can be developed in glass-ceramics, depending upon the chemical composition of the parent glass, and some of the possibilities for titania nucleated aluminosilicate compositions are illustrated by the data given in Table X. The crystal phases quoted for each glass-ceramic type would not necessarily all be present in a given composition but quite often two or more crystal phases may occur. In addition to these phases, rutile or anatase which are crystalline forms of titanium dioxide may also be present in the glass-ceramics.

C. Special Glass-ceramic Processes

1. *Photosensitive Glass-ceramics and Chemical Machining*

Certain types of photosensitively nucleated glass-ceramics can be chemically etched in a controlled process to produce components having precise dimensions. This process is known as chemical machining and it is of sufficient importance to be described in detail.

The chemical machining process was first developed by Stookey (1954) for application to photosensitive glasses. The glasses were of the lithium-alumino-silicate type and contained small amounts of copper, silver or gold as the photosensitive constituents. When these glasses were irradiated with ultra-

TABLE X

CRYSTAL PHASES WHICH OCCUR IN ALUMINO-SILICATE GLASS-CERAMICS

Glass-ceramic type	Crystal phases
$CaO–Al_2O_3–SiO_2–TiO_2$	Anorthite, $CaO\ Al_2O_3\ 2\ SiO_2$ Cristobalite, SiO_2 Sphene, $CaO\ TiO_2\ SiO_2$ Tridymite, SiO_2
$ZnO–Al_2O_3–SiO_2–TiO_2$	Gahnite, $ZnO\ Al_2O_3$ Willemite, $2\ ZnO\ SiO_2$ Cristobalite, SiO_2
$BaO–Al_2O_3–SiO_2–TiO_2$	Dibarium trisilicate, $2\ BaO\ 3\ SiO_2$
$PbO–Al_2O_3–SiO_2–TiO_2$	Lead titanate, $PbOTiO_2$
$MnO–Al_2O_3–SiO_2–TiO_2$	Pyrophanite, $MnO\ TiO_2$ Cristobalite, SiO_2 Manganese cordierite, $2\ MnO\ 2\ Al_2O_3\ 5\ SiO_2$ Mullite, $3\ Al_2O_3\ 2\ SiO_2$ Tridymite, SiO_2 Quartz, SiO_2

violet light through a mask or negative, a latent image was formed in the glass due to the production of atoms of the photosensitive metals in the irradiated regions. During subsequent heat-treatment of the glass, the metal atoms aggregated to form microcrystals and these catalysed the crystallisation of lithium disilicate and other crystals. Thus the effects of irradiation and heat-treatment were to produce an opacified image in the glass. The important discovery was made that the opacified regions of the glass were at least ten times more soluble in dilute hydrofluoric acid than the clear non-irradiated regions. Thus the opacified regions could be dissolved, leaving the clear regions relatively unchanged. By this means intricate patterns could be etched into the glass.

The chemical machining process for glass-ceramics closely resembles that for glasses, especially in the initial stages of the process. British Patent no. 752 243 (Stookey, 1956) gives the compositions of suitable glass-ceramics and these are within the range: SiO_2: 60–85 per cent; Li_2O: 5·5–15 per cent; Al_2O_3:

2–25 per cent. In addition to the essential constituents other oxides, including K_2O, Na_2O or ZnO, may be present in the glass compositions. The materials also contain a photosensitive metal selected from the group: 0·001–0·003 per cent gold computed as Au; 0·001 to 0·03 per cent silver computed as AgCl; and 0·001 to 1·0 per cent copper computed as Cu_2O.

The materials are melted and shaped by standard techniques and the glass articles (usually flat sheets) are irradiated through a suitable mask with ultra-violet light to produce the latent image. The glass is then heat-treated at a temperature between the annealing point (viscosity: $10^{13 \cdot 4}$ poises) and the fibre softening point (viscosity: $10^{7 \cdot 6}$ poises) to cause aggregation of the photosensitive metal atoms into tiny crystals. These subsequently catalyse the crystallisation of a further phase which is usually lithium metasilicate. As was found for the earlier photosensitive glasses, the opacified parts of the glass are much more soluble in dilute hydrofluoric acid than the clear portions so that the glass articles can be selectively etched at this stage to dissolve the regions which were exposed to the ultraviolet radiation.

After etching, the glass article is re-exposed to the ultraviolet radiation to develop photosensitive metal atoms within the previously unexposed regions of the glass. A further heat-treatment process is then applied, first to develop the metallic crystals and subsequently to produce the lithium metasilicate crystals; this is accomplished by heating at a temperature in the region of 500 to 540°C. The temperature is then slowly increased and further crystallisation occurs so that when the maximum temperature of heat-treatment (between 800° and 950°C) is reached the glass has been converted into a substantially polycrystalline ceramic. In addition to lithium metasilicate, lithium disilicate and quartz crystals may be formed in materials of high silica content. For compositions with high alumina contents crystals of beta-spodumene or of a beta-spodumene/quartz solid solution may develop.

This process provides a very convenient way of machining glass-ceramics and the production of plates with very intricate patterns of holes is possible. The etching ratio of the exposed glass to the unexposed glass is about 15 to 1 so that the holes have a taper of approximately 4 degrees. The glass can be etched from one side only giving a conical shaped hole, or from both sides giving a "diabolo" shaped hole. Of course, the etching process can be arrested at any desired stage so that grooves or shallow holes can be produced. Holes up to $\frac{1}{4}$ inch in diameter can be produced to an accuracy within 1 thousandth of an inch and up to 360 000 fine holes can be drilled per square inch.

2. *Surface Crystallisation of Glasses*

While in the strictest sense glasses for which only a relatively thin surface layer is crystallised should not be described as glass-ceramics, the process by which they are produced closely resemble the glass-ceramic process. Also, the crystallisation processes are similar to or identical with those taking place in glass-ceramics. The object of surface crystallisation is to obtain an increase of mechanical strength while causing little or no deterioration of the optical transparency of the glass. In some cases, the enhancement of mechanical strength is achieved by generating a permanent compressive stress in the surface of the glass component. Although bulk glasses are relatively weak in tension, their compressive strengths are extremely high and they rarely, if ever, fail under a stress of this sign. The presence of a surface compressive layer therefore increases the mechanical strength because neutralisation of this stress must first occur before fracture can be initiated. The presence of high compressive stresses in the surface requires the existence of balancing tensile stresses in the interior so that glass strengthened in this way tends to fail "explosively" once the surface compressive layer is penetrated.

It is likely that surface crystallisation of glass can result in higher mechanical strengths by an additional mechanism. This is that the presence of a high volume fraction of very small crystals in a residual glass matrix will result in limitation of the severity of stress-raising flaws. In this case, significant enhancement of strength might be achieved without the necessity for high built-in stresses.

The generation of high compressive stresses in the surface requires the production of crystals of lower thermal expansion coefficient than the uncrystallised interior glass. Olcott and Stookey (1962) showed that the glasses basically of the $Li_2O–Al_2O_3–SiO_2$ type could be surface-crystallised at temperatures in the range 860 to 960°C to produce a surface layer of transparent hexagonal crystals that were isostructural with β-quartz. This layer had a very low coefficient of thermal expansion and thus high compressive stresses were generated in it during cooling. In addition to the major oxide components, the glasses could contain certain minor constituents without adversely affecting the process and titanium dioxide, added in small amounts, actually improved the process. It was shown that there was an optimum thickness range for the surface crystallised layer. This was 60 to 80 μm; excessively long heat-treatments, giving rise to thicker layers, caused disintegration of the glass during cooling while insufficient treatment gave a surface layer that was too thin to withstand abrasion satisfactorily. The moduli of rupture of abraded specimens of the surface crystallised glasses were very high being in the region of 600–700 MNm^{-2} (87 000 to 101 500 lb in^{-2}). The materials were reasonably transparent since the refractive indices of the crystals and the residual glass were fairly closely matched.

In a somewhat different process, Garfinkel *et al.* (1962) produced a similar surface crystallised layer on glasses essentially of the $Na_2O–Al_2O_3–SiO_2$ type containing small concentrations of Li_2O and substantial (5 to 6 per cent) amounts of TiO_2. The glasses were treated in a molten salt bath predominantly comprising lithium sulphate but also containing sodium sulphate. The treatment which was for periods of 5 to 10 minutes at temperatures in the range 860 to 900°C, resulted in the development of a surface crystalline layer comprising a solid solution of β-eucryptite and quartz. The former crystal is known to have a very low coefficient of thermal expansion and the resulting differential contraction therefore caused high surface compressive stresses to be generated. The titania present in the glass was considered to play a vital role in nucleating the crystallisation of the desired phases. By this process, transparent surface-crystallised glasses having strengths as high as 800 MNm^{-2} (116 000 lb in^{-2}) were produced.

McMillan *et al.* (1969b) showed that even glasses of simple compositions could be strengthened by surface crystallisation. A glass of the molecular percentage composition Li_2O 30 SiO_2 70 after heat-treatment for periods of 1 hour at temperatures in the range 500 to 600°C showed strength enhancement when tested in the unabraded condition. Treatment at 575°C, for example, gave a material having a mean rupture modulus of 380 MNm^{-2} (55 000 lb in^{-2}). It was found, however, that abrasion of the specimens before carrying out mechanical strength determinations, caused a marked reduction of strength suggesting that the surface crystallised layer was very thin. Similar but less-marked enhancement of strength occurred for a Li_2O–SiO_2 glass containing 1 molecular per cent P_2O_5 indicating that this oxide tends to inhibit the surface crystallisation process.

Identification of the crystals present in the surface layer proved difficult since X-ray diffraction analysis yielded a very weak pattern. Positive identification was therefore not possible but one strong line a few weak, diffuse lines were consistent with the crystals being lithium disilicate.

A possible explanation of the enhanced strength of the surface-crystallised glass is that the surface layers are stressed in compression owing to differential contraction of the surface layer and the interior. Birefringence measurements did not reveal strain in the glassy core and also generation of such a stress would require the surface layer to have a lower coefficient of thermal expansion than the uncrystallised glass. The thermal expansion of lithium disilicate prepared by fully crystallising a glass of the disilicate composition was 96×10^{-7}°C^{-1} and was thus higher than that of the uncrystallised glass (87×10^{-7}°C^{-1}), therefore ruling out the possibility of surface compressive stresses being developed.

It seems more likely that the enhancement of strength resulted from limitation of the severity of surface flaws (microcracks) by the polycrystalline microstructure developed in the glass surface. This topic will be discussed more fully in connection with a further system of surface-crystallised glasses described below.

Partridge and McMillan (1974) showed that glasses of the ZnO–Al_2O_3–SiO_2 type could be greatly strengthened by the application of a surface crystallisation process. A suitable glass had the weight percentage composition: ZnO 44·7, Al_2O_3 14·0, SiO_2 41·3 and other glasses were derived from this by substituting a variety of other oxides in small percentages for silica. Heat-treatment was generally carried out in air at temperatures in the range 700 to 800°C for periods typically of 24 hours though in certain cases periods of up to 150 hours were employed. During heat-treatment a surface layer of crystals having the keatite structure was produced. For the base glass the optimum thickness of the surface crystallised layer to give the maximum mechanical strength was found to be 70 μm but depending on the nature of the added oxides, optimum layer thicknesses ranged from 30 to 160 μm. It was shown that lightly abrading the glasses either with sand or 220 mesh silicon carbide grit prior to heat-treatment resulted in increased strength and in some cases improved transparency of the surface-crystallised glasses. Microscopy of the specimens demonstrated that the effect of the pre-abrasion was to enhance the density of nucleation sites for surface crystallisation resulting in a more uniform and finer grained microstructure. In a number of cases, pre-abrasion resulted in 50 per cent increase in the mean modulus of rupture of surface-crystallised glass. A further observation was that crystal nucleation was adversely affected if the heat-treatment was carried out in dried "forming gas" (90 per cent N_2: 10 per cent H_2) and the mechanical strength of the product was therefore impaired.

It was possible, using optimum conditions of heat treatment, to produce materials having mean cross-breaking strengths in the range 630 to 830 MNm^{-2} (91 000–120 000 lb in^{-2}) and these high strengths were accompanied by reasonable to good optical transparency.

In considering the reasons for the high strengths of these materials, it should be noted that the stuffed keatite crystal phase is structurally similar to β-spodumene and like the latter has a low coefficient of thermal expansion. Undoubtedly, therefore, the high strength is partly accounted for by the differential contraction of the surface crystalline layer and the uncrystallised glass core. It seems likely, however, that there is another contributory factor since although the glassy interior of the specimens was stressed in tension, the magnitude of the stress as determined by birefringence measurements appeared inadequate to allow the presence of sufficiently high surface compressive stresses to account fully for the strength enhancement.

It was suggested that the microstructure of the surface-crystallised layer resulted in restriction of the sizes of stress raising flaws that could be present and thereby resulted in strength enhancement. Flaws originally present in the glass surface were thought to act as nucleation sites and thereby to be eliminated and the fact that pre-abrasion gave rise to improved nucleation and thereby to better mechanical strength lent support to this idea. It was also deduced that oxygen, either as the gas itself or in the form of water vapour, played a vital role in the nucleation process and that possibly recombination of oxygen with oxygen-deficient sites in the glass surface was involved.

3. *Glass-ceramics having Improved Mechanical Properties*

The strengthening of glasses by ion exchange treatment in which alkali metal ions in the glass surface are replaced by larger ions at a temperature below the strain point has been known for some years (Kiztler, 1961). High compressive stresses are generated in the glass surface during cooling because of the larger volume requirements of the substituted ions.

Similarly, surface compressive stresses can be developed in glass-ceramics and two types of process are possible: those involving simple solid solution and those inducing phase transformations at the surface. Karsetter and Voss (1967) described an example of the first process in which Na^+ ions were exchanged for Li^t ions in the surface of a glass-ceramic containing a β-spodumene solid solution as the major phase. The large surface compressive stresses generated during cooling of the glass-ceramic after ion exchange resulted in a three-fold increase of the modulus of rupture of abraded specimens to 350 MNm^{-2} (51 000 lb in^{-2}). Beall *et al.* (1967) described a further example in which two Li^t ions were exchanged per Mg^{2+} ion in β-quartz solid solutions containing magnesium as an interstitial ion. A two-fold effect resulted from the ion exchange. First, the volume of the unit cell was increased and secondly, the thermal expansion coefficient of the lithium-containing derivative was significantly lower than that of the original crystal. The combined effect was to induce surface compressive stresses resulting in the achievement of a very high strength.

Duke *et al.* (1967) described the application of an ion exchange treatment to bring about a phase transformation in a glass-ceramic surface. Treatment of a glass-ceramic containing nepheline in a potassium salt bath brought about transformation of the crystals in the surface layer to kalsilite. This transformation is accompanied by a 10 per cent increase of volume and therefore a very high surface compressive stress is generated. By this means it was possible to achieve a modulus of rupture of 1400 MNm^{-2} (200 000 lb in^{-2}).

4. *Glass-ceramic Composites*

Although glass-ceramics generally have high mechanical strengths, improvement in fracture toughness is desirable. One way in which this can be achieved is to incorporate high strength fibres or whiskers in a glass-ceramic matrix.

Aveston (1972) has described a cordierite type glass-ceramic reinforced with carbon or silicon carbide fibres. A carbon fibre reinforced material containing 30 volume per cent fibre had a modulus of rupture of 550 MNm^{-2} (80 000 lb in^{-2}) and a work of fracture of $1{\cdot}5 \times 10^4$ Jm^{-2}. In the case of a silicon carbide fibre reinforced material prepared by hot pressing it was found necessary to carry out the process at a temperature above 1250°C to avoid the formation of μ-cordierite which gave rise to undesirable stresses owing to mismatch of thermal expansion coefficients. Using the optimum conditions, a mean strength of 680 MNm^{-2} (99 000 lb in^{-2}) was achieved together with a work of fracture of 2×10^4 Jm^{-2}.

Sambell *et al.* (1972a, b) demonstrated that randomly orientated carbon fibres incorporated in a glass-ceramic of the Li_2O–Al_2O_3–SiO_2 type caused a reduction of mechanical strength but gave a significant increase of fracture toughness. The use of aligned fibres enabled a high strength of 680 MNm^{-2} (99 000 lb in^{-2}) and a work of fracture of 3×10^3 Jm^{-2} to be achieved. The former value represented a three-fold increase and the latter an increase of three orders of magnitude over the values for the unreinforced glass-ceramic. Phillips (1974) also reported high strengths and high values for work of fracture for a carbon fibre reinforced glass-ceramic. It was considered that fibre pull-out was an important process contributing to the work of fracture.

The composites discussed so far have all incorporated brittle fibres but in some applications the use of ductile fibres may be an advantage. Donald and McMillan (1976) developed nickel fibre reinforced glass-ceramics and obtained a remarkably high work of fracture value of $1{\cdot}5 \times 10^4$ Jm^{-2} combined with a flexural strength of 180 MNm^{-2} (26 000 lb in^{-2}). Several processes contributed to the excellent fracture toughness of the nickel reinforced glass-ceramic but fibre pull-out played a major role.

5. *Machinable Glass-ceramics*

Glass-ceramics which combine generally good physical properties with the ability to be machined by near-conventional metal working processes clearly represent a significant advance in materials technology. Beall (1971) and Chyung *et al.* (1974) have described a family of machinable glass-ceramics in which the major crystal phase is fluorophlogopite mica, $KMg_3AlSi_3O_{10}F_2$. These are made by heat-treating glasses having weight percentage compositions in the range SiO_2 30–50, B_2O_3 3–20, Al_2O_3 10–20, K_2O 4–12, MgO 15–25, F 4–10. After a crystallisation heat treatment at 950 to 1050°C a microstructure of interlocking mica crystals as illustrated in Fig. 51 is developed.

The glass-ceramics have good thermal shock resistance even though their thermal expansion coefficients are fairly high (50 to 115×10^{-7}°C^{-1}). They

FIG. 51. Electron micrograph of a machinable glass-ceramic.

are also stable up to 700°C but at higher temperatures fluorine may be lost causing structural degradation. The strengths range from 14 to 140 MNm^{-2} (2000–20 000 lb in^{-2}) and are inversely proportional to the mean mica flake diameter. The fracture surface energies are high because the random orientation and interlocking of the mica crystals causes deflection and blunting of cracks.

The unique microstructure enables the mica glass-ceramics to be machined to close tolerances by drilling, sawing or lathe turning. It is the combination of easy cleavage of the mica flakes and the crack deflection/blunting mechanism that confers machinability.

6. *Glass-ceramics having Orientated Microstructures*

The orientation of crystals within a glass-ceramic is normally random except for crystals growing from surface nucleation sites for which radial alignment often occurs. If a microstructure comprising needle-like crystals exhibiting planar or axial alignment within the bulk of the glass-ceramic component could be generated, beneficial effects upon mechanical strength and fracture toughness might be expected. It would also be anticipated that a glass-ceramic having an aligned microstructure would possess an interesting and possible useful anisotropy of physical properties.

Atkinson and McMillan (1977) have described a hot extrusion technique whereby the major crystal phases in a glass-ceramic can be orientated in a preferred direction. The glass-ceramics employed had molecular percentage compositions of SiO_2 61·0–67·5, Li_2O 24·0–30·5, K_2O 1·5, Al_2O_3 1·0, P_2O_5 1·0, B_2O_3 5·0.

The technique of direct extrusion was used in which a billet of the material, contained in a cylinder, is compressed by a plunger which forces the material through a die aperature. For extrusion, a glass billet was placed in the die and heat treated *in situ* for a pre-determined period of time at a fixed temperature to achieve partial crystallisation. Extrusion temperatures of 800 to 880°C were used and the applied loads ranged from 2500 to 100 kg depending on the extrusion temperature. At each extrusion temperature the glass billet was heat treated for 30 minutes in the die before pressure was applied.

During extrusion a parabolic distribution of velocity, V, is established, owing to frictional forces at the walls of the die, of the form:

$$V = \Delta p/4\eta l\ (r^2 - x^2)$$

where Δp is the pressure difference between the entrance and exit of the die, η is the viscosity, l and r are the length and radius of the die respectively and x is a radial component. The radial velocity gradient thus established causes crystals to rotate during extrusion so that their long axes become aligned with the direction of extrusion.

For the glass-ceramics studied, the crystal phases produced were lithium disilicate in a needle-like habit and tridymite in a more equant form. Excellent alignment of the lithium disilicate crystals was achieved and even the near-spherical tridymite crystals appeared to show partial alignment. The extent of orientation was established by statistical analysis of scanning electron micrographs of the extruded glass-ceramic a typical example of which is shown in Fig. 52. The crystallographic orientation of the lithium disilicate crystals in the extruded glass-ceramic was determined by an X-ray diffraction technique in which the intensity of the (hkl) reflections were measured relative to that from the (111) planes. It was concluded that while the lithium disilicate crystals were crystallographically aligned with the 001 directions parallel to the extrusion axis, the tridymite phase showed no such orientation.

The properties of the extruded glass-ceramic showed interesting differences from those of the same material heat treated in the standard manner. For example the average density which was $2{\cdot}65 \pm 0{\cdot}01$ gm/cc was about 10 per cent higher for the extruded material. The Knoop hardness was found to increase to a maximum when the indenter was aligned with its axis at 90° to the

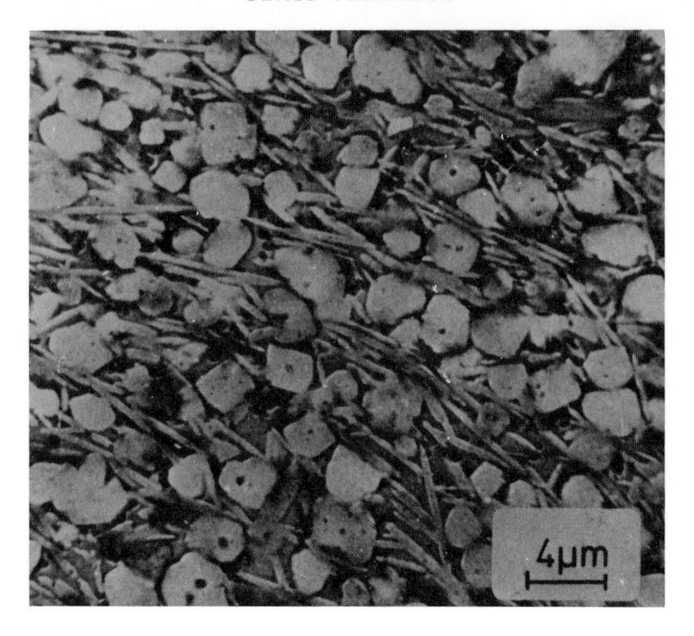

FIG. 52. Electron micrograph of a glass-ceramic having an orientated microstructure.

direction of extrusion. Also the flexural strength of abraded specimens when the material was stressed in the direction parallel to that of extrusion was 250 MNm^{-2} (36 000 lb in^{-2}) and this value was almost twice as large as that in the perpendicular direction. Similar anisotropy was shown by the elastic modulus. It was also found that the electrical conductivity in the extrusion direction was greater than that in a perpendicular direction by a factor varying between 2 and 40.

A method of producing glass-ceramic fibres having an orientated microstructure has been described by Maries and Rogers (1975). In this technique, rods approximately 1 mm diameter of a glass of the weight percentage composition 30 CaO, 20 ZnO, 50 SiO_2 are drawn at a controlled speed through an electrically heated platinum coil. Conditions are adjusted so that the temperature of the molten zone is above the liquidus temperature and crystallisation is initiated by introducing a platinum probe into the melt initially. A stable growth front is thereby established at a position and temperature where the growth rate equals the drawing speed. The crystal phases that can be developed are α- or β-$CaSiO_3$ and the $\alpha \rightarrow \beta$ inversion temperature is above the peak temperature employed (about 1500°C). Low drawing rates (less than 20 $\mu m\ sec^{-1}$) cause the nucleation and growth of α-$CaSiO_3$ while high rates (above 50 $\mu m\ sec^{-1}$) give rise to an uncrystallised glass rod. At intermediate drawing speeds β-$CaSiO_3$ crystals are formed and these take the form of 5 μm diameter fibres comprising practically the whole cross section of the rod and exhibiting very good axial orientation, confirmed by optical microscopy and X-ray diffraction. No significant orientation or fibrous character was found for α-$CaSiO_3$ grown at lower drawing speeds.

The orientated β-$CaSiO_3$ glass-ceramic rods show superior mechanical properties having a rupture strength of 540 MNm^{-2} (78 000 lb in^{-2}) and an elastic modulus of 124 GNm^{-2} (18×10^6 lb in^{-2}). The corresponding values for the unorientated α-$CaSiO_3$ glass-ceramic were 320 MNm^{-2} (46 000 lb in^{-2}) and 91 GNm^{-2} (13×10^6 lb in^{-2}).

Chapter 5

THE PROPERTIES OF GLASS-CERAMICS

Ceramics made by the controlled crystallisation of glass have outstanding mechanical, thermal and electrical properties, and it is useful to compare the properties of glass-ceramics with those of related materials such as glasses or ceramics made by conventional methods. In addition, wherever possible, it is of value to consider how the properties of a glass-ceramic are related to its chemical composition, its crystallographic constitution and microstructure.

A. General Physical and Chemical Properties

1. *Microstructure and Porosity*

One of the notable characteristics of glass-ceramics is the extremely fine grain size and it is likely that this feature is responsible in a large measure for the valuable properties of the materials. It is true to say that a glass-ceramic can have an almost ideal polycrystalline structure since, in addition to the fine texture, the crystals are fairly uniform in size and they are randomly oriented.

In general the average crystal size in useful glass-ceramics is not greater than a few microns and materials with mean crystal sizes as small as 200 to 300 Å are known. For some materials prepared by the crystallisation of glasses, however, greater mean grain sizes are obtained especially where the nucleation density is low and the crystal growth is spherulitic in nature but such materials do not have good mechanical strengths.

In addition to the crystalline phases, there is usually present a residual glass phase. It should be noted that this phase does not normally have the same chemical composition as the parent glass since it will be deficient in those oxides which have taken part in crystal formation.

In contrast to the fine microstructure normally characteristic of glass-ceramics, the average crystal size in sintered alumina ceramics is usually in the vicinity of 10 to 20 microns and that in mineral-based ceramics such as an electrical porcelain may be up to 40 microns.

As will have been seen from the glass-ceramic microstructures illustrated in earlier chapters, considerable variations are encountered in the morphologies of the crystalline and residual glass phases and in their volume fractions. In seeking to explain the properties of glass-ceramics it is necessary to have a reasonably accurate measure of the volume fractions of crystalline and vitreous phases and of the average dimensions of crystals present. For this purpose electron microscopy combined with the well established stereological techniques of lineal or areal analysis or point counting methods as described by Dehoff and Rhines (1968) and Underwood (1970) are utilised. If the compositions, densities and volume fractions of the crystal phases are known, it is possible to compute the chemical composition of the residual glass phase. A knowledge of this can be particularly important in connection with investigations of electrical properties and of chemical stability.

Williams *et al.* (1967) have described additional methods that can be used to determine the volume fraction of crystalline phase. In one technique, after identifying the crystalline phases by X-ray diffraction, a synthetic matching standard is prepared and the crystalline content determined quantitatively from the integrated peak intensity. Another method makes use of differences in solubility in various reagents to separate the crystalline and vitreous phases. This technique cannot be used universally, however, because solvents that will sufficiently differentiate the phases cannot always be found. For low expansion glass-ceramics of the Li_2O–Al_2O_3–SiO_2 type, 0·1 N HF is found to be particularly useful; the reaction is stopped by the addition of a slight excess of boric acid after a few minutes. For this type of glass-ceramic, the glass phase is soluble under these conditions while the crystal phases β-spodumene, rutile and aluminium titanate are not.

Typical results for the low expansion glass-ceramic gave the volume fraction of crystal phases as 0·86 to 0·90 thus the volume fraction of residual glass is in the region of 0·1. For more complex glass-ceramics containing MgO and B_2O_3 in addition to Li_2O, Al_2O_3 and SiO_2, crystallisation was found to be less extensive since in this case a volume fraction of residual glass of 0·3 was found. A BaO–Al_2O_3–SiO_2 glass-ceramic containing mullite as the principal crystalline phase, contained an even higher volume fraction of vitreous phase of about 0·6.

It is clear that considerable variety with regard to the proportions of crystalline and vitreous phases can be encountered in glass-ceramic microstructures. This is true also of the morphology of the phases. Where the volume fraction of crystalline phase is low (less than 0·35 to 0·40) the glass forms an almost continuous matrix containing isolated crystals and it would be expected in this case that the properties of the glass-ceramic would be strongly influenced by those of the vitreous phase. Where the volume fractions of the two phases are roughly equal, the two phases tend to form interpenetrating networks. At high volume fractions of crystalline phase, the glass either takes the form of a thin layer between adjacent crystals or in some cases forms isolated pockets at the grain boundaries. Figure 53 shows the

FIG. 53. The structure of a titania-nucleated magnesia-alumina-silica glass-ceramic.

microstructure of a glass-ceramic in which the volume fraction of glass phase is relatively small. The crystals which comprise a mixture of cordierite and cristobalite, together with a small proportion of rutile, have an average size of 0·5 to 1 micron with a maximum size in the region of 2 microns. It would be expected that such a microstructure would confer high mechanical strength and this is in fact found to be the case.

Since the crystals in a glass-ceramic are formed by precipitation from a homogeneous liquid (glass) phase under conditions where there are usually no stresses that can cause alignment, they are randomly orientated. (As pointed out in chapter 4 orientated microstructures can be achieved by suitable techniques.) The absence of orientation means that the properties of a glass-ceramic are isotropic. Even though some of the crystals present may have different properties in different crystallographic directions (see p. 184) these are averaged out in the glass-ceramic.

Although ceramics made by conventional techniques can often be quite impermeable to liquids or gases so that their apparent porosity is zero and they are vacuum tight, they are rarely, if ever, completely free from closed pores. In a sintered high-alumina ceramic the percentage closed porosity is usually in the region of 5 to 10 per cent and in a felspathic porcelain of the type used for electrical insulators it is about 10 per cent. In contrast with conventional ceramics, glass-ceramics are entirely free from any type of porosity, providing the glass from which they are prepared is free from gas bubbles. Porosity does not develop during the conversion from the glass to ceramic state since the overall volume changes are very small. Quite often the volume change is a shrinkage but even when the conversion is accompanied by a volume increase, voids do not develop within the interior of the material, the volume increase being due to the production of crystals of lower density than the original glass. The complete absence of pores in glass-ceramics is a characteristic which favours the development of good properties since pores will reduce the mechanical strength by diminishing the useful cross-section of the material and in addition they may act as flaws at which internal fractures could originate. Certain electrical properties may be adversely affected by the presence of pores and, of course, pores would also reduce the thermal conductivity; this may or may not be desirable, depending upon the application for which the material is to be used.

While the dense, fine texture of glass-ceramics serves to distinguish them from conventional ceramics and forms the basis which determines many of their special characteristics, the nature and proportions of the crystal phases present are extremely important in deciding the individual characteristics of a glass-ceramic. Consideration of the effects of different crystal types upon properties will be given in sections dealing with specific properties.

2. *Density*

The density of glass-ceramics lie within a similar range to those for glasses or conventional ceramics, as is shown by the data given in Table XI. The density of a glass-ceramic will be an additive function of the densities of the various crystal and glass phases present. Since the volume change which occurs during conversion from the glass state to the glass-ceramic is usually small, it would be expected that the effects of various oxides upon the densities of glass-ceramics would be similar to those observed with conventional glasses. Broadly speaking, this is true because those oxides, such as barium or lead oxides, which tend to confer high densities upon glasses also result in glass-ceramics with high densities. Similarly, glass-ceramics having lithia as a major constituent have low densities, as do glasses containing this oxide. It is also found that increases of the proportions of MgO, CaO, ZnO, BaO or PbO at the expenses of Al_2O_3 or SiO_2 in the glass-ceramics leads to higher densities and that BaO and PbO exert the most marked effects.

Although the general level of the density values for a particular type of glass-ceramic is determined by the overall chemical composition, the

TABLE XI

DENSITIES OF GLASS-CERAMICS, GLASSES AND CONVENTIONAL CERAMICS

Material	Range of densities (g/cm^3)
*Glass-ceramics**	
Li_2O–Al_2O_3–SiO_2–TiO_2	2·42 to 2·57
MgO–Al_2O_3–SiO_2–TiO_2	2·49 to 2·68
CaO–Al_2O_3–SiO_2–TiO_2	2·48 to 2·80
ZnO–Al_2O_3–SiO_2–TiO_2	2·99 to 3·13
BaO–Al_2O_3–SiO_2–TiO_2	2·96 to 5·88
PbO–Al_2O_3–SiO_2–TiO_2	3·50 to 5·76
Glasses	
Fused silica	2·20
Soda-lime-silica	2·40 to 2·55
Low expansion borosilicate	2·23
Potash-soda-lead-silica	2·85 to 4·00
High lead, alkali free	5·4 to 6·2
Ceramics	
High tension porcelain	2·3 to 2·5
Steatite ceramics	2·5 to 2·7
Forsterite ceramics	2·7 to 2·8
High alumina ceramics	3·4 to 4·0

* Data extracted from British Patent no. 829 447 (Stookey, 1960) and Australian Patent no. 46 230 (Stookey, 1959b).

constitution of the crystal and glass phases present will determine the densities of specific glass-ceramics. For instance, silica could be present as a constituent of the glass phase, as free silica in the form of cristobalite or quartz, or as a constituent of a complex silicate crystal, and its contribution to the density of the glass-ceramic could be different in all these forms. Quartz, for example, has an appreciably higher density (2·65 g/cm^3) than cristobalite (2·32 g/cm^3) so that a change in the relative amounts of these two crystals could have a marked effect on the density of a glass-ceramic. This has been demonstrated for a glass-ceramic which can be heat-treated to produce the free silica entirely as cristobalite or quartz or having intermediate proportions of these crystals present. It was found that the density of the glass-ceramic increased from 2·46 g/cm^3 when the silica was present entirely as cristobalite to 2·59 g/cm^3 when the silica was present entirely as quartz. Since these glass-ceramics were produced from the same parent glass, it is clear that the volume changes occurring during the crystallisation process can be noticeably influenced by the form in which the free silica appears. The effects of variations of the heat-treatment schedule upon the density of the glass-ceramics are also shown by data for glass ceramics containing either beta-spodumene or beta-eucryptite as major phases. For heat-treatments to maximum temperatures of 800 to 900°C as compared with temperatures of 1000 to 1200°C it was found that the densities of the glass-ceramics generally differed by about 1·6 per cent, although in one case a difference of density of 2·8 per cent occurred. Usually the higher heat-treatment temperature favoured the formation of beta-spodumene as the major crystalline phase as opposed to beta-eucryptite and the lower densities were generally associated with the higher heat-treatment, although this was not invariably the case.

In some glass-ceramics, oxides present in minor amounts may exert a significant influence on the density because they affect the types of major crystal phases which are present. For example, it has been shown that when zinc oxide is substituted on a molecular basis for aluminium oxide in phosphate-catalysed glass-ceramics of the Li_2O–Al_2O_3–SiO_2 type, the first additions of ZnO cause only a small increase of the density. When the molecular ratio of ZnO exceeds that of Al_2O_3, however, the density increases quite rapidly. This effect is due to the presence of quartz as a major crystal phase since this crystal only appears when the molecular proportion of ZnO exceeds that of Al_2O_3.

3. *Chemical Durability*

The resistance of a material to chemical attack by water or other reagents is of considerable practical importance. The processes involved are often very complex and methods of studying chemical durability usually involve subjecting a sample of carefully controlled surface area to closely specified test conditions. Often, the process of attack is accelerated, as compared with anticipated service conditions, by employing conditions of increased temperature and, in some cases, pressure.

Tests of this nature have indicated that glass-ceramics in general possess good chemical stability and that they compare favourably in this respect with other ceramic-type materials.

In many cases, it is likely that when a glass-ceramic is chemically attacked the initial effect is upon the glass phase present. This occurs because the early stages of attack involve ion exchange between hydrogen (or hydroxonium ions) and mobile cations (usually alkali metal ions) in the glass. Subsequently the silica network structure can be attacked by a process of hydration. The greater mobility of alkali metal ions in the glass phase as compared with that of similar ions incorporated in crystal phases, will lead to greater reactivity of the glass phase and hence to inferior resistance to chemical attack.

The achievement of high chemical durability in glass-ceramics therefore requires the volume of residual glass phase to be small and also that the chemical composition of this phase favours good stability. The avoidance of high concentrations of alkali metal oxides in the glass phase will assist in attaining improved chemical durabilities. A silica-rich glass phase containing in addition alumina and zinc oxide together with alkaline earth oxides will favour the attainment of good chemical stability.

These general effects have been confirmed for glass-ceramics basically of the lithium silicate type containing minor amounts (5 per cent or so) of oxides such as Al_2O_3, MgO and ZnO. The tests were conducted by determining the losses in weight which occurred when standard "grit" samples of the glass-ceramics were subjected to the action of boiling water for 2 hours. This is a severe test because of the large surface area which is exposed to chemical attack. Increase in the lithia content had the most marked effect on the chemical durability. For example, the extent of chemical attack was increased by a factor of fourteen when the lithia content was raised from about 12 weight per cent to about 20 weight per cent. These results were obtained for glass-ceramics having relatively simple compositions and possessing rather high lithia contents, so that they are not necessarily applicable to all types of glass-ceramics. It was shown, however, that a more complex glass-ceramic having a low lithia content (2·5 weight per cent) and containing in addition substantial amounts of alumina and magnesia was very durable, since in this case the loss of weight during the "grit" test was less than one-tenth of that observed for glass-ceramics containing about 12 weight per cent of lithia.

Kay and Doremus (1974) reported on the durability of a glass-ceramic containing beta-spodumene crystals. This material was only slightly susceptible to stress-enhanced reduction of strength by reaction with water at 23°C. This result was in contrast to effects observed for glasses, crystalline oxides and other glass-ceramics. It was suggested that the apparently high stability of the beta-spodumene glass-ceramic was a result of the low reactivity of the crystalline phase.

Further evidence regarding the stability of glass-ceramics against attack by water has been obtained in tests in which specimens were exposed to conditions of 100 per cent relative humidity at 70°C for periods up to 5000 hours. A lithium-zinc-silicate glass-ceramic showed no evidence of surface deterioration after this test, while soda-lime-silica glass was severely weathered and had a thick surface deposit after the same treatment. Even borosilicate glass of the type used for chemical laboratory ware showed some signs of weathering in this test, since a surface "bloom" was developed. On this basis, the chemical resistance of glass-ceramics would appear to compare favourably with those of glasses which are widely used under conditions demanding good chemical stability.

Certain types of glass-ceramics have good resistance to attack by corrosive chemical reagents. Low expansion glass-ceramics derived from lithium-aluminosilicate glasses are only slightly inferior to borosilicate chemically resistant glass with regard to attack by strong acids and are somewhat more resistant to attack by alkaline solutions. It has been reported from the Soviet Union (*Anon*, 1962) that a thermal shock resistant glass-ceramic (possibly based on a lithium-aluminosilicate glass) has a high resistance to chemical attack. It was stated that a mixture of nitric and sulphuric acid can remain in vessels made of this material for 1000 hours without causing deterioration and also that even the most harmful chlorine compounds do not produce a noticeable effect after contact for two days at a temperature of 300°C. Glass-ceramics derived from magnesium-alumino-silicate glass compositions are slightly less resistant to attack by strong acids and alkalis than are chemically resistant boro-silicate glasses. In this connexion, it has been reported (British Patent no. 903 706, 1962) that a glass-ceramic of this type withstood submersion in concentrated hydrochloric acid for two days without suffering appreciable etching. Similarly, exposure to hydrofluoric acid fumes for 5 to 10 minutes did not cause appreciable etching.

Even at high temperatures, certain types of glass-ceramic can be resistant to attack by corrosive gases. It has been shown in the author's laboratory, for example, that lithium-alumino-silicate glass-ceramics of medium thermal expansion coefficient are quite unaffected by exposure to chlorine or hydrogen chloride gases for a period of six hours at a temperature of 800°C.

B. Mechanical Properties

1. *Mechanical Strength*

a. General considerations. The mechanical strength of a material is one of the most important properties since it is often the major factor in determining the suitability of the material for a particular application. In addition, other important characteristics, such as the ability of a material to withstand sudden temperature changes without failure, are strongly influenced by the mechanical strength. Not only is it desirable to have a high mechanical strength at normal ambient temperatures but also to retain high strengths at elevated temperatures since quite often the material may be required to operate under such conditions. For these reasons it is necessary to know how the strength of a glass-ceramic is influenced both by its constitution and by external factors.

Before examining the mechanical strengths of glass-ceramics in detail there are certain general aspects of mechanical failure which must be considered. At room temperature glass-ceramics, like ordinary glasses and ceramics, are brittle materials. This means that they exhibit no region of ductility or plasticity and that they show perfectly elastic behaviour up to the load which causes failure. Like other brittle solids, they have relatively high elesticities and they are capable of developing forked fractures. Important experimental observations with glasses are that mechanical strengths measured can vary widely from specimen to specimen and also that the strength is markedly dependent upon the specimen size. For example, in a batch of apparently identical glass specimens, the coefficient of variation of the experimental results may be as high as 25 to 30 per cent. In addition, while freshly drawn, glass fibres may have tensile strengths of the order of 7000 MNm^{-2} (10^6 lb in^{-2}), the strength measured for more massive glass specimens would rarely exceed 70 MNm^{-2} (10 000 lb in^{-2}). Both of these effects are explainable in terms of the hypothesis proposed by Griffith (1920), which suggests that glass specimens are riddled with cracks or flaws of varying severity and that these cracks act as stress multipliers. The probability of a severe flaw having the correct orientation to influence the mechanical strength will diminish as the specimen size decreases, so that fine glass fibres would be expected to be stronger than more massive specimens. Similarly, the variability of mechanical test values for a particular set of specimens is explainable in terms of the presence or absence of flaws of varying severity in critical orientations for the specimen under test. If the number of flaws present is small, such as may occur for freshly drawn glass rod or for rods which have been lightly etched in hydrofluoric acid, very high mechanical strength values can be obtained, and even for 0·5 cm diameter rods cross-breaking strengths of 1400 MNm^{-2} ($\sim$ 200 000 lb in^{-2}) or higher have been measured. Conversely, abrasion of the glass surface, which introduces severe flaws, lowers the mechanical strength but at the same time it reduces the variability of the test results since now all specimens contain large numbers of flaws of comparable severity. For this reason, many experimenters prefer to carry out investigations of mechanical strengths using specimens which have been subjected to standard abrading procedures.

The foregoing observations, which have been made largely for ordinary glasses, are also relevant to the study of glass-ceramics since these materials have a very fine microstructure and it is to be expected that surface flaws could influence the mechanical strength. In connection with this it should be noted that, when comparisons are made between mechanical strengths of different types of glass-ceramic and other materials, the conditions of test must be taken into account, if at all possible. Sometimes this is difficult because different laboratories use different sizes of test specimen and, while some investigators use standard abrasion procedures others use unabraded specimens, believing that the test results give a closer approach to the "practical" strength of the material even though individual test values may vary appreciably.

Interpretations of the results of mechanical strength investigations on glasses have relied very heavily on the well known expression for mechanical strength, σ, derived by Griffith (1920):

$$\sigma = \sqrt{\frac{2E\gamma}{\pi c}}$$

where E is the elastic modulus, γ is the fracture surface energy and c is the length of the critical flaw. From this it is evident that increase of strength would occur only if E or γ were increased or if c were reduced. It is of interest to consider the relative importance of these parameters with regard to the mechanical strength of glass-ceramics. Before doing this, however, it will be helpful to review the experimental evidence.

b. The mechanical strengths of various glass-ceramic types. As a general rule the strengths of glass-ceramics are high compared with ordinary glasses and with other types of ceramics. The data given in Table XII provide a basis for comparison but it should be noted that the actual value obtained for a given material is greatly influenced by the surface condition of the test specimens. The values quoted are for specimens that have received no special treatment.

Table XIII gives ranges of values reported for various glass-ceramic systems. It will be noted that the strengths of glass ceramics can vary widely

TABLE XII

MODULUS OF RUPTURE VALUES FOR GLASS-CERAMICS AND OTHER MATERIALS

(The values are those typically obtained for 0·5 cm diameter rods in 3-point loading)

Material	Range of values	
	MNm^{-2}	$lb\ in^{-2}$
Glasses	55–70	8000–10 000
Glass-ceramics	70–350	10 000–50 000
Electrical porcelain (unglazed)	70–80	10 000–12 000
Electrical porcelain (glazed)	80–140	12 000–20 000
High alumina ceramic (95 per cent Al_2O_3)	200–350	30 000–50 000

TABLE XIII

MODULI OF RUPTURE OF VARIOUS TYPES OF GLASS-CERAMICS

Glass-ceramic system	Nucleating agent	Moduli of rupture MNm^{-2}	Moduli of rupture $lb\,in^{-2}$	Ref.
Li_2O–SiO_2	None	30–50	4000–7000	1
Li_2O–SiO_2	P_2O_5	110–398	16 000–58 000	1,2
Li_2O–ZnO–SiO_2	P_2O_5	176–340	25 000–49 000	3
Li_2O–Al_2O_3–SiO_2	TiO_2	112–122	16 000–18 000	4
MgO–Al_2O_3–SiO_2	TiO_2	119–259	17 000–38 000	4
CaO–Al_2O_3–SiO_2	TiO_2	120	17 000	4
BaO–Al_2O_3–SiO_2	TiO_2	55–64	8 000–9 000	4
ZnO–Al_2O_3–SiO_2	TiO_2	38–131	6000–19 000	4
ZnO–MgO–Al_2O_3–SiO_2	ZrO_2	69–103	10 000–15 000	5
Na_2O–Al_2O_3–SiO_2	TiO_2	84	12 000	6
Na_20–BaO–Al_2O_3–SiO_2	TiO_2	89–114	13 000–16 000	6

References: 1. McMillan (1971), 2. Hing and McMillan (1973b), 3. McMillan *et al.* (1966a), 4. Stookey (1960), 5. Beall (1972), 6. Duke *et al.* (1968).

depending upon the glass-ceramic system and also on the heat treatment cycle employed. Some general observations can be made concerning the data in Table XIII. First, the importance of the nucleating agent in promoting development of the type of microstructure necessary to achieve good mechanical strengths is apparent from the first two sets of data. Inclusion of P_2O_5 in Li_2O–SiO_2 glass-ceramics enables a greater than seven-fold increase of strength to be achieved. In these materials lithium disilicate is present as a major crystalline phase and this is also true for glass-ceramics of the Li_2O–ZnO–SiO_2 system for which high strengths can be achieved. Glass-ceramics of the MgO–Al_2O_3–SiO_2 system containing cordierite as a major phase can also exhibit high mechanical strengths. One reason for this may be that the mismatch in thermal expansion co-efficient between the crystalline phase and the residual glass phase is small. Hence microstresses within the glass-ceramic can be relatively low. The lithium-aluminosilicate materials are of great interest because it is from this system that materials of low thermal expansion coefficient and high thermal shock resistance are derived. The mechanical strengths of these materials are generally lower than those of the cordierite glass-ceramics. This lower strength is probably due to the presence of significant internal stresses in the glass-ceramics. The presence of beta-spodumene, which has a very low coefficient of thermal expansion, is likely to result in circumferential tensile stresses in the residual glass phase around the crystals. Even if these stresses do not cause microcracking they are likely to lower the overall strength of the glass-ceramic. If beta-eucryptite is present as a major phase, the situation is worsened because this crystal in the form of a polycrystalline aggregate has a highly negative coefficient of thermal expansion. Stookey (1959b) showed that while glass-ceramics containing beta-spodumene gave bending strengths in the range 129–143 MNm^{-2} (18 000–20 000 lb in^{-2}), materials containing beta-eucryptite had strengths of 70–79 MNm^{-2} (10 000–11 000 lb in^{-2}). The marked anisotropy of beta-eucryptite is also likely to affect adversely the strengths of glass-ceramics containing this crystal (this effect is discussed on p. 184).

It should be noted that glass-ceramics of the Li_2O–Al_2O_3–SiO_2 type in which the crystals having very low thermal expansion coefficients are not developed can have high mechanical strengths. Such materials have low alumina contents and the major crystal phases are usually lithium disilicate and quartz or cristobalite. McMillan and Partridge (1963a) reported a material of this type having a rupture modulus of 286 MNm^{-2} (40 000 lb in^{-2}). The absence of low expansion phases was confirmed by the thermal expansion coefficient of the glass-ceramic which was $102 \times 10^{-7}\,°C^{-1}$.

Few systematic studies have been reported concerning the effects of changing the parent glass composition, (and hence the nature of the crystal phases developed) upon the mechanical strengths of glass-ceramics. In one such investigation the effects of replacing Al_2O_3 by ZnO were studied for a glass of the molecular percentage composition: $SiO_2 + Li_2O + K_2O$: 91·5; P_2O_5: 1·0; Al_2O_3: $(7{\cdot}5 - x)$; ZnO: x. As the results given in Table XIV show the moduli of rupture of the glass-ceramics increase significantly as the ZnO content is increased.

The increase of strength is quite small for the first additions of zinc oxide but a fairly sharp increase occurs when the molecular proportion of ZnO exceeds that of Al_2O_3. This effect is most likely to be connected with changes in the proportions and types of the crystal phases which are present as the composition is varied. In the compositions where the Al_2O_3 contents exceeds the ZnO content, the major phase is a silica-rich beta-spodumene solid solution of the approximate composition Li_2O Al_2O_3 $10SiO_2$. Lithium disilicate (Li_2O $2SiO_2$) is also present, the amount increasing as the Al_2O_3 content diminishes. When the ZnO content becomes greater than the Al_2O_3 content, beta-spodumene is no longer detected and the major phase is quartz; a large proportion of lithium disilicate is also present. Thus in glass-ceramics of this type, it is clear that the disappearance of beta-spodumene and the appearance of quartz, as the major crystal phase is associated with an increase of mechanical strength. In these materials it appears that the presence of the low expansion phase, beta-spodumene, may result in the generation of unfavourable internal stresses and its replacement by higher expansion phases leads to enhanced mechanical strength.

TABLE XIV

EFFECTS OF ZINC OXIDE AND ALUMINIUM OXIDE CONTENTS ON THE PROPERTIES OF GLASS-CERAMICS

Mol. per cent ZnO	Mol. per cent Al_2O_3	Modulus of rupture		Modulus of elasticity (lb/in^2)		Thermal expansion coefficient
		MN^{-2}	lb in^{-2}	10^4 MNm^{-2}	10^6 lb in^{-2}	[× 10^7 (20–500°C)]
0	7·5	138	20 000	not measured		50·4
1·5	6·0	152	22 000	7·0	10·2	52·5
3·0	4·5	152	22 000	9·7	14·1	62·1
4·5	3·0	193	28 000	9·8	14·2	108·6
6·0	1·5	203	29 500	12·1	17·6	118·4
7·5	0	272	39 500	14·9	21·6	126·6

c. Influence of heat-treatment schedule upon mechanical strength. Systematic studies of the effects of variation of the heat-treatment cycle in modifying the microstructure and mineralogical constitution of glass-ceramics are of value in providing a basis for understanding the factors determining the strength of the materials. In this case, of course, the chemical compositions of the glass-ceramics are unchanged but large variations in the crystallographic constitution can occur and appreciable effects upon physical properties, such as strength, may result.

Work by Watanabe *et al.* (1962) on a composition essentially of the Li_2O–Al_2O_3–SiO_2 type but containing a small proportion of K_2O and using metallic silver as the nucleating agent showed that the mechanical strength was strongly influenced by the heat-treatment process. For rods abraded before heat-treatment the strength increased progressively as the maximum heat-treatment temperature was raised from 500° to 900°C. Specimens heat treated at the latter temperature showed a more than two-fold increase of strength over that of the parent glass.

The influence of heat-treatment upon the mechanical properties of Li_2O–SiO_2 glass-ceramics has been reported by McMillan (1971). Figure 54 summarises the results of modulus of rupture measurements on abraded rod specimens that have been nucleated for 1 hour at temperatures in the range of 450 to 575°C, followed by a standard crystallisation treatment of 1 hour at 750°C. The P_2O_5 containing glass produced stronger glass-ceramics for all heat-treatments and this was a result of the much finer microstructure (grain sizes 8·5 to 50 μm) as compared with those for the P_2O_5-free glass (43 to 250

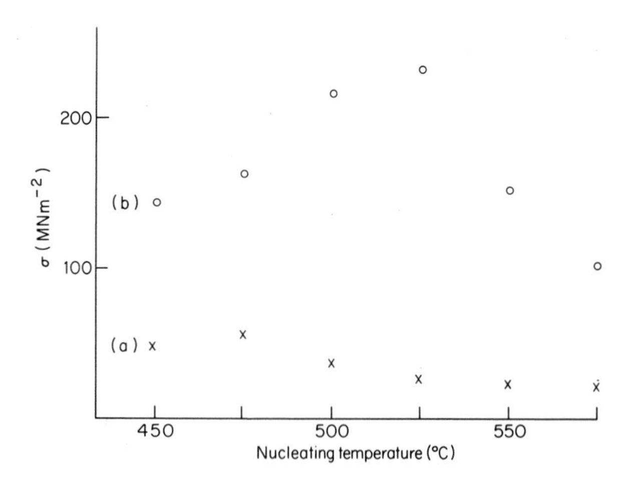

FIG. 54. Modulus of rupture as a function of nucleation temperature for (a) 70 SiO_2, 30 Li_2O glass-ceramic, (b) 69 SiO_2; 30 Li_2O; 1 P_2O_5 glass-ceramic (molecular percentages).

μm). In the case of the glass containing P_2O_5 the results clearly demonstrate the importance of nucleation treatment in the development of the optimum mechanical strength. For the P_2O_5 free glass, the mechanical strengths achieved were not greatly influenced by the nucleation treatment.

McMillan *et al.* (1966b) found that the development of lithium disilicate crystals by the heat-treatment of a glass of the molecular percentage composition Li_2O 22·3, K_2O 1·5, ZnO 3·0, SiO_2 72·5, P_2O_5 0·7, resulted in a significant increase of strength. Specimens of the glass were given a nucleation treatment of 500°C for 1 hour followed by heat-treatment at temperatures in the range of 600 to 850°C for 1 hour. The results, summarised in Fig. 55 show that the mechanical strength is practically doubled by heat-treatment at 600°C. X-ray diffraction analysis and electron microscopy showed that the glass-ceramic produced by this heat-treatment consisted of approximately equal proportions of lithium disilicate and uncrystallised glass, though the crystalline phase was not well formed. After heat-treatment at 700°C, the proportions of crystalline and glass phases were still approximately equal though the characteristic dendrites of the disilicate now appeared clearly defined. Heat-treatment at temperatures above 700°C resulted in

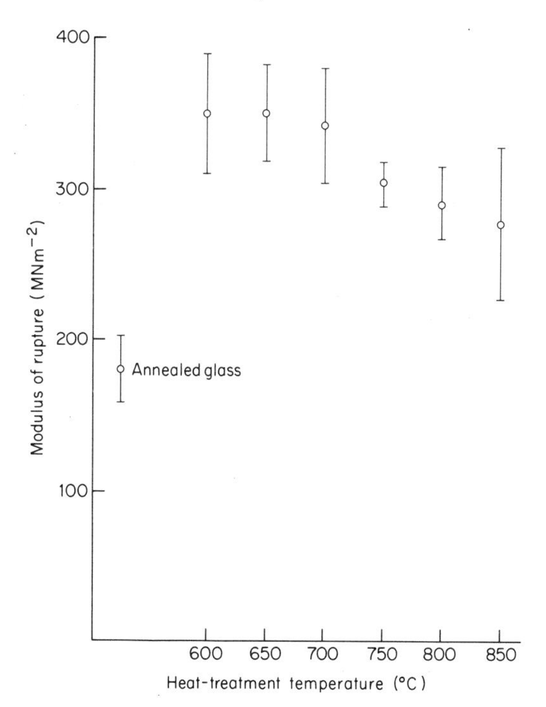

FIG. 55. Influence of heat-treatment temperature on the modulus of rupture of a glass-ceramic.

crystallisation of the residual glass and the development of increasing proportions of crystalline silica (usually quartz) until the glass-ceramic heat-treated at 850°C contained approximately equal proportions of lithium disilicate and crystalline silica with about 5 per cent of residual glass phase. The formation of the crystalline silica phase was accompanied by a progressive reduction of the mechanical strength of the glass-ceramic. It is likely that this effect is related to the generation of unfavourable microstresses in the glass-ceramic. The crystalline forms of silica found to occur in these glass-ceramics (quartz and cristobalite) both have very much higher thermal expansion coefficient than lithium disilicate or the residual glass phase. Therefore, differential contraction during cooling of the glass-ceramic will cause the generation of significant radial tensile stresses associated with the silica crystals. In the extreme, microcracking could result but even if this did not occur, the presence of high internal tensile stresses would be most likely to cause weakening of the glass-ceramic.

A useful study of the mechanical properties of glass-ceramics of the modified Li_2O–SiO_2 type has been reported by Borom *et al.* (1975) who studied the effects of varying both nucleation and growth treatments on the moduli of rupture of glass-ceramics derived from a glass having the molecular percentage composition: Li_2O 23·7, K_2O 2·8, Al_2O_3 2·8, SiO_2 67·1, B_2O_3 2·6, P_2O_5 1·0. Nucleation temperatures of 580° and 645°C. were employed in conjunction with growth temperatures in the range 715° to 978°C. The highest strength of 399 MNm^{-2} (55 800 lb in^{-2}) was obtained for a glass-ceramic which had received a nucleation treatment of 1 hour at 645°C and a growth treatment of 168 hours at 830°C. Quantitative X-ray diffraction was employed to evaluate the crystal contents of the various glass-ceramics. All of the materials contained lithium phosphate Li_3PO_4 in amounts ranging from 1·5 to 5 weight per cent. The most striking feature, however, was the change in the type of lithium silicate crystal present. For the lower growth temperatures and shorter growth times lithium metasilicate was present in significant amounts while for higher temperature treatments lithium disilicate became the predominant phase. It was noticeable that the glass-ceramics containing the metasilicate phase were significantly weaker than those containing the disilicate phase. Also, it appeared to be significant that the strongest group of specimens contained the maximum concentration of lithium disilicate crystals (46 weight per cent) and only a trace of the metasilicate phase. Borom *et al.* pointed out that the strengthening mechanism in their glass-ceramics could not simply be due to an increase of elastic modulus. The very strong material showed a more than three-fold increase of rupture modulus over that of the parent glass but the elastic modulus was increased only by a factor of 1·3. The suggestion was made that the much larger thermal expansion mismatch between the lithium metasilicate crystals and the residual glass phase as compared with the situation existing in the disilicate glass-ceramics was responsible for the observed strength differences. The thermal expansion coefficients for the various phases were: residual glass $90 \times 10^{-7}\,^{\circ}C^{-1}$; lithium disilicate $114 \times 10^{-7}\,^{\circ}C^{-1}$; lithium metasilicate $130 \times 10^{-7}\,^{\circ}C^{-1}$. Calculation showed that the radial tensile stresses at the crystal-glass interface would be greater for the lithium metasilicate glass-ceramics by a factor of three. Thus an increased possibility arises of microcracking at the interface causing reduction of mechanical strength.

Microstructural effects upon the mechanical strengths of glass-ceramics were also shown by the work of Hing and McMillan (1973b) who investigated the effects of heat-treatment on the properties of a glass-ceramic of the molecular percentage composition Li_2O 29, K_2O 1, ZnO 1, SiO_2 68, P_2O_5 1. Specimens of this were given a nucleation treatment of 1 hour at 500°C followed by treatment at 550°C for periods of 1 to 48 hours to accomplish crystallisation. The crystal phase developed was lithium disilicate throughout. Increase of the duration of heat treatment from 1 hour to 24 hours caused a steady increase of rupture modulus from 176 MNm^{-2} (25 500 lb in^{-2}) to 343 MNm^{-2} (49 700 lb in^{-2}). Thereafter strength reduction occurred and the value obtained for specimens heat treated for 48 hours was 239 MNm^{-2} (34 600 lb in^{-2}). These changes in strength were considered to be related to microstructural changes taking place. The initial increase of strength was attributed to an increase of the volume fraction of the lithium disilicate crystals. At the longer heat-treatment time, little further change in the volume fraction occurred but the mean crystal size increased as a result of a coarsening process. It was considered that the mechanical strength results were explainable in terms of the variation of a microstructural parameter, λ, the mean free path in the residual glass phase. This parameter is given by

$$\lambda = d(1 - Vf)/Vf$$

were d is the mean crystal size and Vf is the volume fraction of the crystal phase. The relationship shown in Fig. 56 was found to exist. It will be seen that for values of λ less than about 8 μm the strength is inversely related to $\lambda^{\frac{1}{2}}$. For higher values of λ the strength is independent of the variation of this parameter. A possible explanation of these observations is that the dispersion of lithium disilicate crystals in the residual glass phase serves to limit the dimensions of flaws that can be present in the glass. When, however, the mean inter-crystal spacing exceeds the length of flaws that exist in the uncrystallised glass, limitation of flaw size is no longer possible and the strength will be independent of the particle spacing. Freiman and Hench (1972) had earlier reported that the strength of Li_2O–SiO_2 glass-ceramics increased as the inter-crystal spacing decreased. Their materials, however, showed extensive

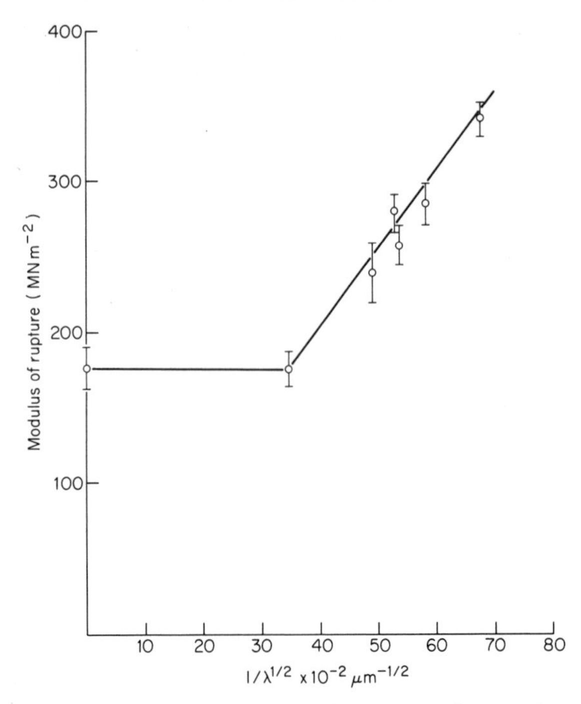

FIG. 56. Relationship between modulus of rupture and mean free path, λ, for a glass-ceramic.

microcracking and the strength was more likely to be controlled by this than by microflaws inherent to the glass-ceramics.

The mechanical properties of glass-ceramics derived from the composition 3BaO $5SiO_2$ were studied by Freiman *et al.* (1974). The specimens were given a nucleation treatment at 700°C followed by heat-treatment for periods of up to 48 hours at temperatures in the range 775 to 1300°C. Three types of microstructure were observed depending on the heat-treatment. For temperatures up to 825°C and crystallisation times of 1 hour, a spherulitic morphology developed. Prolonged heat-treatment at 850°C caused secondary crystallisation of residual glass between the arms of the spherulites and at the boundaries between the grains. Heat-treatment at temperatures of 1000°C and upwards led to the development of lath-like crystals. Flexural strength measurements and determinations of fracture energy using the double cantilever beam technique were made. It was found that the mechanical strength was increased when secondary crystallisation occurred. The strongest group of specimens, having a mean flexural strength of 161 MNm^{-2} (23 300 lb in^{-2}), was that heat-treated at 1000°C for 24 hours to develop the lath-like morphology. Heat-treatment at higher temperatures caused a drastic fall of strength and this was attributed to the occurrence of microcracking. The authors concluded that the increases of strength observed for the lower temperature heat-treatments were due to increases of elastic modulus and fracture energy. They based this conclusion on the observation that a plot of σ versus $\sqrt{E\gamma}$ gave a straight line passing through the origin. The implication of this result is that the crack length, c, in the Griffith equation remained constant and independent of microstructural changes. A flaw size of 54 μm was estimated from the slope of the curve.

In considering whether this result might be held to be generally applicable to glass-ceramics, it should be borne in mind that the method of abrasion used in the preparation of the specimens may be very important. For any brittle material, there will exist the possibility of introducing surface microcracks of a size that over-rides any microstructural constraints if the abrasion treatment is of sufficient severity. If the material, in addition contains microstresses (as evidenced for some of the $BaO–SiO_2$ glass-ceramics by the occurrence of internal microcracking) the possibility of introducing severe surface microcracks during abrasion will be enhanced. Under conditions of severe abrasion, therefore, the surface microcracks may be of an approximately constant depth that is more characteristic of the abrading method than of the material being examined. In the case of the $BaO–SiO_2$ glass-ceramics, roughly cut bars were subjected to the crystallisation treatment, ground to the final shape and polished to a 20 μm finish. It is uncertain whether the polishing treatment would be adequate to remove deep flaws introduced during the grinding operations.

d. Effects of surface condition upon mechanical strength. So far we have dealt with those factors affecting the mechanical strength which may be thought of as being intrinsic to the glass-ceramic. There are other factors, however, which are not directly connected with the internal structure of the materials but which can strongly influence the mechanical strength. The effect of the condition of the surface upon the mechanical strength has already been mentioned, and this factor justifies closer examination. Watanabe *et al.* (1962) in the investigation already referred to, showed that the glass-ceramic which had been heat-treated to achieve maximum strength was weakened by a standard abrasion treatment and that the abraded strength was 81 per cent of the unabraded strength. For the same abrasion treatment, however, glass which was not heat-treated showed a much greater reduction of strength since the abraded specimens had strengths which were only 14 per cent of those of unabraded specimens. These results suggest that glass-ceramics are much less susceptible to surface damage than are glasses and this may be one of the reasons for the higher strengths of glass-ceramics. An interesting observation made by Watanabe *et al.* (1962) was that if very special care was taken to avoid surface damage to the glass before heat-treatment, the effect of the crystallisation heat-treatment was to reduce the strength rather than to increase it. This result is in direct contrast to that observed for glass rods which had been abraded before heat-treatment. From this it appears that while the crystallisation heat-treatment raises the "practical" strength of glass (since under ordinary conditions glass would always suffer surface damage due to abrasion or other causes) it does not give greater strengths than can be obtained with glass having "pristine" surfaces for which, of course, very high mechanical strength values can be measured.

Experiments conducted with a lithium-zinc-silicate glass-ceramic by Phillips (1962) showed that the mechanical strength was reduced to about 80 per cent of its original value by a fairly severe abrasion treatment. With the same treatment, a commercial borosilicate glass (Pyrex) suffered a reduction of strength to 45 per cent of the unabraded value. Etching of the unabraded glass-ceramic with hydrofluoric acid (a procedure which often strengthens ordinary glass quite markedly) had little effect and certainly did not result in a gain of strength. For the same glass-ceramic, however, surface treatment carried out on the glass before the crystallisation heat-treatment showed interesting effects with regard to the strength of the final material. In these experiments some of the glass rods were etched for a brief period in concentrated hydrofluoric acid under carefully controlled conditions, other rods were abraded by rolling them in a ball mill for 10 minutes with 80 grit silicon carbide, while other rods received no special surface treatment.

TABLE XV

EFFECTS OF SURFACE TREATMENT BEFORE CRYSTALLISATION UPON THE STRENGTH OF A GLASS-CERAMIC

Group no.	Surface treatment	Crystallisation heat-treatment	Strength factor
1	None given	None given	1·0
2	None given	Normal schedule	2·4
3	Etched in HF	None given	3·3
4	Etched in HF	Normal schedule	3·4
5	Abraded	Normal schedule	1·4

Strength measurements were carried out on rods having the various surface treatments using specimens which had received no heat-treatment (ie were in the glass form) and specimens which had received the standard crystallisation heat-treatment. The results of these tests are summarised in Table XV and are expressed for convenience in the form of a strength factor which is defined as the ratio of the mean strength of the particular group of specimens to that of the specimens which received no surface treatment and no crystallisation heat-treatment.

Based on these results the following observations can be made:

(*i*) Etching of the unheat-treated glass increases its mechanical strength by a factor of more than three times (Groups 1 and 3).

(*ii*) The glass-ceramic prepared from etched glass is appreciably stronger than that prepared from unetched glass (Groups 2 and 4) although heat-treatment of etched glass does not increase its strength significantly (Groups 3 and 4).

(*iii*) Abrasion of the glass before heat-treatment results in a material of appreciably lower strength than that produced from unabraded glass (Groups 2 and 5).

Since these experiments were confined to one type of glass-ceramic it would be unwise to assume that they are universally applicable to glass-ceramics. Nevertheless, they provide a guide to some of the factors which must be taken into account when the mechanical properties of these materials are being studied.

A study by Kay and Doremus (1974) has provided additional information concerning the influence of surface conditions. These investigators studied a low expansion glass-ceramic of the titania-nucleated $Li_2O–Al_2O_3–SiO_2$ type in which the main crystalline phase was a beta-spodumene/silica solid solution together with small proportions of rutile and titanate crystals. Measurements of modulus of rupture in 4-point loading were made both at room temperature and liquid nitrogen temperature.

It was found, somewhat surprisingly, that controlled abrasion using 15 μm silicon carbide paper did not cause a reduction of strength. In fact the mean strength at liquid nitrogen temperature appeared to increase by about 8 per cent and this was thought to be statistically significant. A possible explanation of this result is that the abrasion treatment removes a layer from the glass-ceramic surface that adversely affects the strength. A surface layer having a higher overall thermal expansion coefficient than the bulk material would be stressed in tension, and its removal could result in strength enhancement. Certainly this effect has been observed in the author's laboratory for other low expansion glass-ceramics.

Kay and Doremus also showed that etching of the glass-ceramic in 4 per cent HF resulted in significant strength enhancement both at room and liquid nitrogen temperatures. An etch depth of about 3 to 4 μm appeared to give optimum results and on the basis of this it was considered that the un-etched glass-ceramic contained flaws of about 3 μm in depth.

e. Effect of temperature upon mechanical strength. An important variable in the process of mechanical failure is temperature, and it is of both practical and fundamental interest to know the relationship between strength of a glass-ceramic and the temperature. Figure 57 shows the influence of temperature upon the mechanical strength of a commercial glass-ceramic. This material is basically a magnesia-alumina-silica composition utilising titanium dioxide as a nucleation catalyst. Cordierite is present as a major crystalline phase and a crystalline form of silica (cristobalite) is also present. The strength values are for abraded rods which have subsequently received a chemical strengthening treatment. The modulus of rupture at room temperature for this glass-ceramic is high but the strength falls as the temperature increases, the curve having an S-shaped form. Even at 800°C, however, the strength is significantly higher than that of most glasses at room temperature and, of course, at 800°C the strength of ordinary glasses would be negligible since they would undergo viscous flow. At the upper limit of the measurements (1200°C) a useful strength value is still retained. The strength of other types of glass-ceramics has also been observed to fall with increase of temperature, as the results for a lithium-aluminosilicate material of intermediate expansion coefficient given in Fig. 58 (curve A) show. In some cases, however, there is an initial fall of strength, followed by a recovery at higher temperatures. This effect is clearly

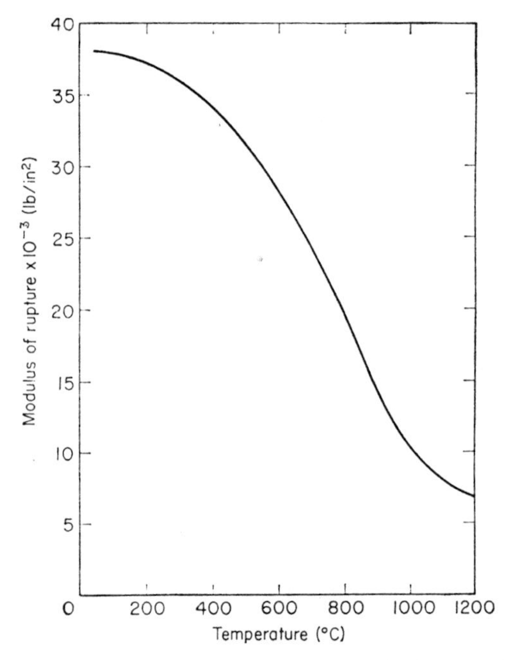

FIG. 57. The effect of temperature upon the mechanical strength of glass-ceramic no. 9606.

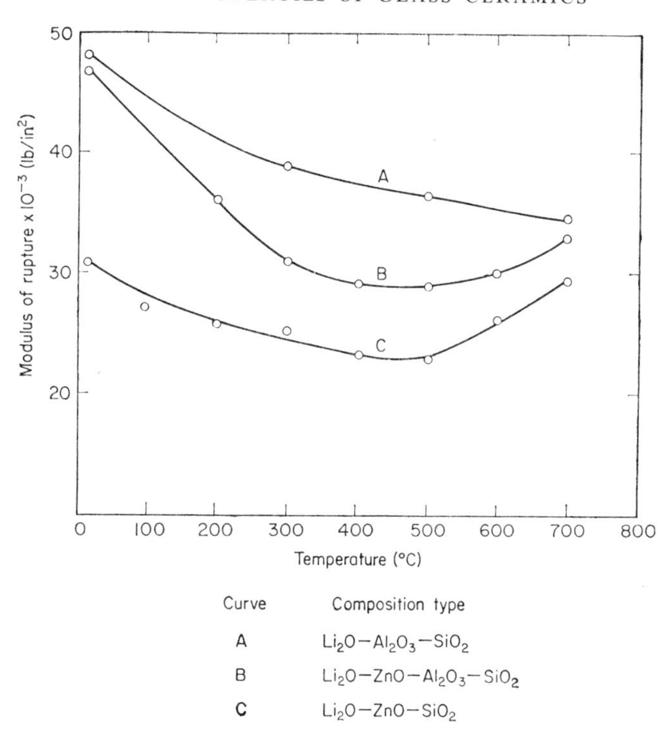

FIG. 58. The effect of temperature upon the moduli of rupture of glass-ceramics.

shown for a lithium-zinc-aluminosilicate composition (curve B). At still higher temperatures, of course, the strength of these materials would diminish once more, since softening or even melting of the materials would occur. In comparative tests, a 95 per cent alumina ceramic also appeared to show a minimum value of strength at 400 to 500°C and several investigators have reported that soda-lime-silica glass exhibits a minimum value of strength at about 200°C. In the case of glass, the existence of a minimum strength may arise from two opposing effects which occur as the temperature increases. One effect is that the intrinsic strength falls due to the weakening of atomic bonding forces. At the same time, however, flaws which are present in the glass surface may become less severe due to a "rounding off" process resulting from diffusion of ions in the glass, and this would tend to increase the measured strength. The combination of the two effects would give rise to a minimum strength at a particular temperature. Whether or not this explanation can be put forward in the case of glass-ceramics is uncertain and further experimental work is necessary to confirm the presence of a minimum in the strength-temperature relationship. In a polyphase material such as a glass-ceramic, it is likely that stresses arising due to differential thermal expansion of crystalline and glassy phases will influence the form of the strength-temperature relationship and the occurrence of a minimum strength at a temperature where the microstresses attained a maximum value would be expected. Borom (1977) reported the temperature dependence of strength for two glass-ceramics of the Li_2O–Al_2O_3–SiO_2 type which contained lithium disilicate as the major crystal phase. Both materials showed a reduction of strength with increase of temperature. For one material, the strength continued to fall as the temperature was increased to a maximum of 550°C, but the other glass-ceramic attained a minimum strength at about 350°C with some recovery as the temperature was increased further. Borom pointed out that the fall of strength could not be attributed to a reduction of the elastic modulus with increased temperature. In the temperature range of interest this parameter decreased by no more than 5 per cent while the mechanical strengths decreased by 30 to 50 per cent. It was suggested that the reduction of strength was caused by the reduction of favourable internal stresses owing to differential expansion of the crystal and glass phases.

f. Tensile, compressive and impact strengths of glass-ceramics. The discussion of mechanical strength of glass-ceramics has largely been confined to considerations of modulus of rupture measurements. This method of measurement has been widely used since it possesses distinct advantages such as the use of easily prepared specimens, and in addition, the method can give reliable mean values without too great a scatter in the results. There are, of course, other strength parameters which it may be desirable to know such as tensile strength, compressive strength and impact strength. Measurement of tensile strength for a brittle material is extremely difficult because stress concentrations arise at the points where the specimen is gripped for applying the tensile load even in carefully conducted tests. Consequently specimens often fail outside the gauge length. For this reason tensile strength measurements on glass-ceramics are felt to be unreliable. In comparative tests where the inaccuracies were probably of the same order, it was shown that the tensile strength of a glass-ceramic was greater than that of a high-tension electrical porcelain by a factor of roughly three. Thus the tensile strengths of the two materials were in approximately the same ratio as their moduli of rupture.

Since glass-ceramics (like glasses) probably always fail in tension, compressive strength has little meaning. In any test designed to measure compressive strength, it is probable that failure occurs in regions where tensile forces are operating, and the shape and size of the specimen, together with the method of applying the load, will strongly influence the compressive stresses necessary to cause fracture. For these reasons, compressive strength values which are quoted can often be quite misleading. However, in a test in which cylindrical specimens $\frac{3}{8}$-inch diameter and $\frac{3}{8}$-inch long were compressively loaded via their end faces, the "compressive strengths" of a glass-ceramic and a 95 per cent alumina ceramic were roughly the same ratio of their moduli of rupture.

The resistance of a material to catastrophic failure during impact can be of great importance in determining its usefulness in engineering applications. For this reason, there is considerable interest in measuring the impact strength of materials. This, however, is not a fundamental quantity but is a complex property determined by a number of other properties. Undoubtedly a high tensile strength contributes towards a good impact resistance but the elastic modulus should not be too high since if the material is too stiff, little energy can be absorbed by deflection of the object. The work of fracture should be high so that if a fracture is initiated, energy is rapidly consumed in creating the new surfaces and the fracture propagation is rapidly brought to a halt. In some cases, failure due to impact may occur at a point some distance from the actual point of impact. This is due to the initiation of elastic waves by the impact which, being reflected by the surfaces of the test piece or component, may become focussed or concentrated in some region. For example, it is not uncommon to observe that rectangular glass plates struck by high velocity liquid drops suffer damage owing to the corners breaking away in addition to the localised damage at the point of impact.

In addition to the complexities of impact failure itself, it should be pointed out that the design of a material for the achievement of high resistance to impact failure requires certain compromises. As discussed earlier, attainment of high tensile strength requires a fine-grained microstructure but such a structure is not the best for optimising impact resistance. The latter requires high fracture toughness and this will not be favoured by a very fine-grained structure. Rather, a coarser-grained structure in which the fracture is deflected around glass-crystal boundaries is needed. Also, the attenuation of elastic shock waves is more likely in a coarse-grained material. Possibly, these apparently conflicting demands can best be achieved by reinforcing the glass-ceramic with a second, usually fibrous phase of significantly different elastic properties from the glass-ceramic matrix, to give the types of composite material already discussed in chapter 4.

Because of the complex nature of impact failure, it is often considered that the only realistic appraisal of a material for a given application is to test components either under actual service conditions or under conditions simulating these as closely as possible. Nevertheless, there is value in attempting to quantify the impact resistance of materials in a standard test firstly with the aim of gaining an understanding of the interactions of more fundamental properties that govern impact failure and secondly to provide a basis for classifying materials with regard to their possible use in engineering applications.

The usual type of test employed is one in which a specimen of circular or rectangular cross-section is struck by a pendulum under standard conditions. The energy loss of the pendulum is taken to correspond to that absorbed by the impact and to give a measure of the impact strength. A Charpy type of test carried out on unnotched glass-ceramic rod specimens 1 cm diameter and 8 cm long at 20°C gave an impact strength of 0·54 ft lb this being roughly three times greater than the value for a 95 per cent alumina ceramic. In this case, the tensile strengths of the two materials were comparable and it is thought that the work of fracture values would not differ greatly. The inferior performance of the alumina ceramic can therefore probably be attributed to the higher elastic modulus (roughly three times that of the glass-ceramic). The impact strength of the glass-ceramic was found to fall to 0·48 ft lb at 300°C and 0·18 ft lb at 500°C.

In an attempt to gain a better understanding of impact testing both from a theoretical and experimental viewpoint McMillan and Tesh (1975) investigated both the experimental techniques itself and the impact failure of several materials. Using an instrumented pendulum equipment and computer calculations they were able to show that complex dynamic effects occurred during impact as a result of vibrations induced in the test specimen.

Some of their results for a glass, a glass-ceramic and two ceramics are summarised in Table XVI. In this case the resistances to impact failure are expressed as the energy loss of the pendulum. It will be seen that in the tests, carried out under approximately the same conditions in each case, the glass-

TABLE XVI

RESULTS OF IMPACT TESTS ON 4 mm DIAMETER RODS

Material	Surface condition	Energy absorbed (mJ)	Modulus of rupture (MNm^{-2})
Soda-lime-silica glass	Abraded, dry	9·3 (30)	42 (10)
Soda-lime-silica glass	Abraded, wet	8·4 (16)	37 (10)
Glass-ceramic	Ground	23 (20)	138 (8)
95% Alumina ceramic	Ground	21 (16)	208 (5)
Electrical porcelain	Ground	15·1 (6)	122 (5)

N.B. Figures in parenthesis are the coefficients of variation (per cent).

ceramic gave the highest value closely followed by the 95 per cent alumina ceramic. Both materials were significantly better than the glass. Table XVI also gives modulus of rupture values measured for specimens of the same dimensions as those used for impact studies.

In order to provide a basis of comparison, calculation was made of the ratio between the energy absorbed during impact γ_I and the strain energy γ_s, stored in the modulus of rupture specimens immediately before failure. It was shown that the strain energy is given by

$$\gamma_s = \pi \sigma_s^2 d^2 l / 48E$$

where σ_s is the modulus of rupture, E is the Young's modulus, d is the specimen diameter and $2l$ the separation between the specimen supports.

It was found for the glass specimens, including the results of other tests not listed in Table XVI, that the ratio γ_I/γ_s was approximately constant at about 4. Also the value for the glass-ceramic specimens (4·3) and for the porcelain specimens (3·2) did not depart greatly from this value. The ratio for the alumina ceramic was, however, higher being 6·2. This is somewhat surprising since the value of σ_s for this material is considerably greater than for the other materials although this is counterbalanced to some extent by a higher elastic modulus. The implication of these results is that the alumina ceramic performed less well in the impact test than would have been expected from its other properties. It should be pointed out that the values for γ_I/γ_s for all of the materials were higher than the ratio of about 2 reported by Haward (1944) for glass. A possible explanation is that the modulus of rupture values obtained by McMillan and Tesh were low owing to the use of a low loading rate in the tests which would allow static fatigue effects to become significant. It seems unlikely that the high value for the ratio γ_I/γ_s found for the alumina ceramic could however be explained in terms of accentuated static fatigue sensitivity for this material.

Fracture toughness, which is closely related to impact strength, will be enhanced if the fracture is compelled to follow a tortuous path through the material. This requires intercrystalline rather than transgranular fracture.

Two factors will influence the mode of fracture. Firstly, if the crystal phases present do not have well-defined cleavage planes, a fracture approaching the glass-crystal boundary may propagate fairly readily through the crystal. On the other hand, when randomly oriented crystals having well-defined cleavage planes are present, the probability is low that they will be aligned in the optimum direction for crack propagation across the glass-crystal boundary and the fracture will therefore tend to be deflected. Hence intercrystalline fracture will be favoured in this case. In a glass-ceramic containing lithium disilicate and quartz as major phases it was observed that fractures generally followed the grain boundaries in the case of the former crystal but crossed the boundary in the case of the latter. This is thought to result from the presence of well-defined cleavage planes in lithium disilicate while this is not the case for quartz. Another example is given by mica-containing glass-ceramics. These crystals, of course, exhibit a very marked cleavage direction and this favours intergranular rather than transgranular fracture.

The second important factor relates to the stresses present at the glass-crystal boundaries. Where the thermal expansion coefficient of the crystal is lower than that of the glass, the radial stresses in both phases are compressive; the circumferential stresses are tensile in the glass and compressive in the crystal. A crack approaching the crystal will thus tend to propagate across the interface because the stresses normal to the glass-crystal interface are compressive. Hence fracture will be transgranular.

When the situation is reversed so that the crystal phase has the higher expansion coefficient, the radial stresses in both phases are compressive. The circumferential stresses are compressive in the glass and tensile in the crystal. Thus intergranular fracture is favoured.

McCollister and Conrad (1966) have confirmed these ideas by investigating fracture propagation in two glass-ceramics having similar geometric microstructures. In the first which was of the Li_2O–Al_2O_3–SiO_2 type, the major crystal phase was a low expansion stuffed quartz phase. The expansion coefficient of this ($6 \times 10^{-7}\,°C^{-1}$) was very much less than that of the glass phase ($41 \times 10^{-7}\,°C^{-1}$) and a mean stress of 117 MNm^{-2} (17 000 lb in^{-2})$\pm$15 per cent was measured at the glass-crystal interface by birefringence. Transgranular fracture predominated as expected. In the second glass-ceramic, 25 μm crystals of calcium fluoride were present in a glass matrix and the thermal expansion coefficient of the former ($195 \times 10^{-7}\,°C^{-1}$) was very much higher than that of the latter ($93 \times 10^{-7}\,°C^{-1}$). The tensile stress at the glass-crystal boundary was $29 >$ MNm^{-2} (43 000 lb in^{-2})°15 per cent and as predicted, the fracture was almost exclusively intergranular.

The implication of these results is that low thermal expansion glass-ceramics for which the residual glass phase is likely to have a higher expansion than the crystal phases present, will exhibit lower fracture toughness than medium to high thermal expansion glass-ceramics.

g. Static fatigue for glass-ceramics. A factor which must be borne in mind when undertaking mechanical strength studies on brittle materials is the possibility of static fatigue. Silicate glasses and oxide ceramics exhibit this effect since the measured strength depends upon the length of time that the load is applied or upon the loading rate. Although these materials can withstand high stresses for a limited period of time, failure will occur at considerably lower stresses if the application of the load is prolonged for a sufficient time. As a result, the rate of loading during the mechanical strength test must be controlled and is usually very low. Also, of course, the existence of static fatigue means that a safety factor (usually at least 2) must be employed in converting mechanical test values for a ceramic to a practical "long-term" strength value which can safely be used in design calculations. Since glass-ceramics are closely related to glasses and conventional ceramics, it would not be surprising to find that they too exhibit static fatigue.

Recent work has shown for glasses that static fatigue involves a species of stress corrosion and that atmospheric water plays a significant role. Charles and Hillig (1965) postulated that under the influence of an applied stress the rate of corrosion at the flaw tip is enhanced. This leads to sharpening of the flaw tip and thereby increases the stress multiplication factor caused by the flaw. At some point the local stress at the flaw tip will exceed the breaking stress of the glass and "fatigue failure" will occur. The corrosion process, it is surmised, first involves exchange of hydrogen ions for alkali metal ions in the glass R^+ (glass) $+ H^+$ (water) $\rightarrow H^+$ (glass) $+ R^+$ (water). The rate of this reaction will be controlled by the diffusion rates of R^+ and H^+ ions in the glass and may be dominated by the mobility of the alkali ions. The network structure of the glass can also be attacked by water according to $\equiv Si–O–Si \equiv + H_2O \rightarrow \equiv Si–HO \ldots . HO–Si \equiv$. It is this reaction which is probably responsible for the delayed failure or static fatigue of glass. Although fused silica glass is attacked only slowly at room temperature, in alkali silicate glasses the reaction takes place at a significant rate. This is not simply due to the formation of an alkaline solution because the reaction proceeds at an enhanced rate even if the pH of the solution in contact with the glass is maintained constant—by adding acid for example.

It is probable that similar reactions can take place with glass-ceramics though for those materials where the alkali metal ions are largely incorporated in crystal phases the susceptibility to static fatigue would be likely to be less than for a glass of the same composition. Watanabe *et al.* (1962) demonstrated the occurrence of static fatigue for a lithium-aluminosilicate glass-ceramic. More recently Kay and Doremus (1974) investigated the behaviour of a glass-ceramic containing beta-spodumene as the major crystalline phase. In the usual method for investigating static fatigue, specimens are loaded with some fraction of the breaking stress observed at liquid nitrogen temperature and the average time to failure is determined. For glasses and ceramics a stress of $\sim 0{\cdot}7$ times the liquid nitrogen failure stress is employed. In the case of the glass-ceramic, however, no specimens failed at 0·8 times this stress in 10^5 seconds. It was only at stresses close to the mean liquid nitrogen fracture stress that all of the specimens failed in less than 10^4 seconds.

It was concluded from this study that the glass-ceramic was significantly less susceptible to static fatigue effects than soda-lime glass, fused silica and polycrystalline alumina. The explanation proposed for this unique behaviour was that the beta-spodumene crystals which comprised 95 per cent of the glass-ceramic exhibit a low reactivity with water. It was suggested that although ion exchange of hydrogen for lithium takes place, the second reaction involving hydration of the silica lattice may be less probable thus reducing the severity of stress corrosion effects.

h. The microstructural basis for the strengths of glass-ceramics. The mechanical strengths of glass-ceramics are higher than those of many conventional ceramics and glasses when comparison is made using specimens that have received the same abrasion treatment. The complete absence of pores has already been mentioned as factor contributing to high strength. This, however, does not provide a sufficient explanation for the observed strengths and it seems that a combination of factors may be responsible. The various possibilities are as follows:

1. The elastic moduli of glass-ceramics (8 to 15×10^4 MNm^{-2}) are significantly greater than those of glasses ($\sim 6 \times 10^4$ MNm^{-2}) and thus theories of mechanical strength based on a critical strain concept would predict a higher strength for glass-ceramics. This explanation is insufficient, however, since glass-ceramics are often stronger than glasses by a greater factor than the ratio of the elastic moduli.
2. Glass-ceramics are harder and more abrasion resistant than glasses and surface damage leading to loss of strength is likely to be less severe for a given abrading condition.
3. The combination of crystal and glass phases having different thermal expansion may lead to a system of microstresses that favours high mechanical strengths. Where the crystal has a higher thermal expansion coefficient than the surrounding glass, the radial stresses in the glass will be tensile and the circumferential stresses compressive. Where the crystal possesses the lower expansion coefficient, the signs of the stresses are reversed. Experience suggests that second system of stresses is less favourable since glass-ceramics containing low expansion crystal phases tend to be weaker than those containing crystals of high expansion.

The presence of anisotropic crystal phases may also be responsible for the generation of unfavourable boundary stresses between the crystal and glass phases. Such phases include quartz, β-eucryptite, corundum and aluminium titanate and many others; the data in Table XVII illustrate the magnitude of the anisotropy for certain crystals that can occur in glass-ceramics. In the extreme, the stresses generated can be high enough to result in cracking and separation between individual grains which will cause a catastrophic reduction of mechanical strength. For a glass-ceramic, it is probable that the boundary stresses are minimised owing to the very small average crystal sizes, but the existence of these stresses is one of the reasons for dependence of strength upon grain size.

TABLE XVII

THERMAL EXPANSION DATA FOR ANISOTROPIC CRYSTALS

Crystal	Thermal expansion coefficients
Quartz	78×10^{-7} parallel to c axis 144×10^{-7} perpendicular to c axis
β-Eucryptite	-176×10^{-7} parallel to c axis $82{\cdot}1 \times 10^{-7}$ perpendicular to c axis
Corundum	$66{\cdot}6 \times 10^{-7}$ parallel to c axis 50×10^{-7} perpendicular to c axis
Aluminimum titanate	118×10^{-7} parallel to a axis 194×10^{-7} parallel to b axis -26×10^{-7} parallel to c axis

4. The presence of crystals in a glass-ceramic causes deflection and possibly blunting of the fracture tip. Thus the work of fracture is increased and the crack may be slowed down or even arrested as it traverses or crosses boundaries between crystalline or glass phases whereas in glass there will be an uninterrupted fracture path. Figures 59 and 60 show the marked differences in fracture surfaces for a glass and a glass ceramic. For the latter, the fracture surface is much rougher since the fracture has generally followed the grain boundaries between crystals. These microstructural effects on fracture propagation are likley to be important in impact failure.

5. It is probable that the mechanical strengths of glass-ceramics are controlled by the severity and distribution of microcracks in the surface. These factors will be controlled by the surface microstructure of the glass-ceramics if, as seems likely, the microcracks are propagated across crystal-glass boundaries with greater difficulty than through the intervening glass. In this case the lengths of microcracks will be related to the sizes and volume fractions of the crystal phases and it is probable that an optimum microstructure will exist which will result in a minimum average microcrack length and therefore in maximum mechanical strength.

Utsumi and Sakka (1970) suggested that the mechanical strength σ, of a glass-ceramic is related to the mean grain diameter, d, by the relationship.

$$\sigma = Kd^{\frac{1}{2}}$$

where K is a constant. The implication of this relationship is that the crack length, c, in the Griffith equation

$$\sigma = \sqrt{\frac{2E\gamma}{\pi c}}$$

is proportional to the grain diameter. Another possibility, however, is that in a glass-crystal composite containing relatively strong crystalline dispersions, the initiation of fracture will occur within the glass matrix. At a sufficiently high volume fraction of the dispersed phase the maximum size of flaw that can exist may be restricted as suggested by Hasselman and Fulrath (1966).

The essential feature proposed is that flaws present in the glass are terminated at the crystal-glass boundaries and the average separation between crystals, the mean free path, will be a critical parameter. The mean free path, λ, is given by

$$\lambda = d(1 - V_f)/V_f$$

where V_f is the volume fraction of the crystalline phase.

If the maximum dimension of the flaws in a glass-ceramic is restricted to λ, the expression for mechanical strength can be written.

$$\sigma = K^1 \lambda^{-\frac{1}{2}}$$

The value of K^1 in this expression is given by $(2E\gamma/\pi)^{\frac{1}{2}}$. If this quantity remains relatively invariant as a function of microstructure, a plot of σ *vs*$\lambda^{-\frac{1}{2}}$ will give a straight line when the value of λ is less than the critical flaw size in uncrystallised glass. This was shown to be so for one glass-ceramic system by Hing and McMillan (1973b) as discussed earlier.

The foregoing discussion does not take into account the effects of microstresses in the glass-ceramic but these could play a large part in modifying the influence of microstructural changes on mechanical strength. In cases where the crystal phases have lower thermal expansion coefficients than the residual glass phase, circumferential tensile stresses and radial compressive stresses will exist in the glass phase immediately surrounding the crystals. Where the sign of the expansion mismatch is reversed, the signs of the stresses will of course be reversed. Thus if the crystalline phases play a part in restricting flaw sizes in the glass phase, flaws having their major axes aligned in the radial direction are likely to be more critical in the former case and the flaw length will be controlled by the inter-crystal spacing ie by λ. In the second case, where high thermal expansion coefficient crystals are present, flaws aligned in the circumferential direction are likely to be more critical. The lengths of such flaws will be proportional to the crystal diameter d and the strength would then be a function of $d^{-\frac{1}{2}}$ rather than $\lambda^{-\frac{1}{2}}$. Practical glass-ceramics may contain a mixture of crystals having thermal expansion coefficients both lower and higher than that of the residual glass-phase and in such cases, the microstructural dependence of strength will be complex.

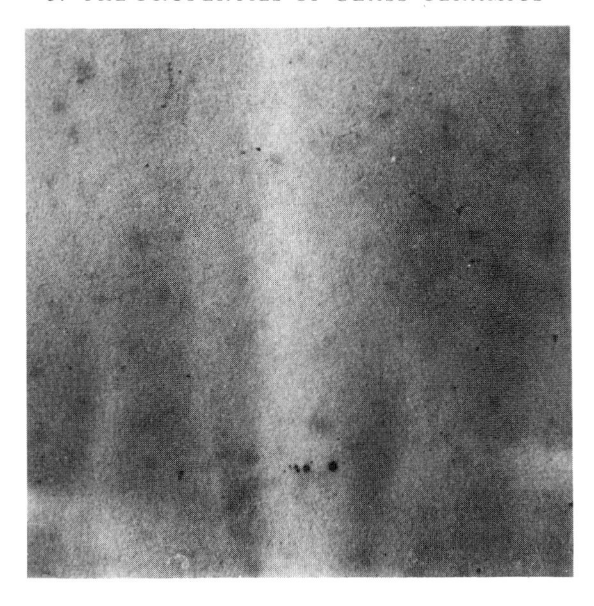

FIG. 59. Fracture surface of a glass before devitrification. (× 10 000)

FIG. 60. Fracture surface of a glass-ceramic. (× 10 000)

The effects of stresses in dispersion-strengthened glasses and glass-ceramics upon the mechanical strengths have been considered by Borom (1977). He concluded that in the case of glass strengthened by the inclusion of Al_2O_3 particles, the increase of strength was a result of the increase of elastic modulus as the volume fraction of the dispersed phase increased. The increases of strength, however, were less than would be predicted from the observed increases of elastic modulus and Borom suggested that the introduction of the Al_2O_3 particles cause an increase rather than a reduction of flaw size. In considering results from a study of the mechanical properties of phosphate-nucleated $Li_2O–Al_2O_3–SiO_2$ glass-ceramics Borom found, however, that by controlled crystallisation the strengths were increased by a factor three or four times greater than could be explained by the increases of elastic moduli. To explain this, it was proposed that a favourable system of microstresses exists in the glass-ceramics such that the residual glass phase is stressed in compression. Essentially the argument was that in a glass-ceramic containing a crystal phase of higher thermal expansion coefficient than the matrix glass phase, the radial tensile stresses in the glass phase around the crystals would decrease to zero at the outer surface of the specimens whereas the circumferential compressive stresses would not. Also interactions between the stress fields of adjacent particles could cause reduction or even reversal of sign of the radial stress component.

It would appear that this argument might suggest that in glass-ceramics containing crystals of lower thermal expansion than the matrix glass, the interfacial stresses would result in weakening of the glass-ceramics as compared with the parent glass. While it is true that glass-ceramics containing low expansion crystal phases generally do not have such high mechanical strengths as those containing high expansion phases, the former materials can be significantly stronger than the glasses from which they are derived.

2. *Elastic Properties*

a. General considerations. The elastic properties of a material are of great importance in determining its behaviour when it is subjected to deformation. For example, the modulus of elasticity fixes the levels of stress which are generated if a glass-ceramic is strained by the application of a temperature gradient. This can occur when the material is suddenly heated or cooled and in this case a low modulus of elasticity is desirable. Similarly, if the glass-ceramic is to be sealed to another material having a different coefficient of thermal expansion (a metal, for example) a low modulus of elasticity is useful since for a given strain the stress will be lower. There are cases, of course, where a high modulus of elasticity may be desirable. This situation exists for glass used in fibre form for reinforcing plastics.

The elastic constants of solids are:
Modulus of elasticity (Young's), E
Modulus of rigidity or torsion, G
Bulk modulus, K
Poisson's ratio, ν
For isotropic bodies the following relationships exist between these constants:

$$G = E/2(1+\nu)$$

$$K = E/3(1-2\nu)$$

$$\nu = E/2G-1$$

b. Elastic constants for glass-ceramics. The moduli of elasticity for glass-ceramics are higher than those of ordinary glasses and of some conventional ceramics, but they are lower than those of sintered pure oxide ceramics as the data in Table XVIII show.

TABLE XVIII

MODULI OF ELASTICITY FOR GLASS-CERAMICS AND OTHER MATERIALS

Materials	Range of values of E 10^4 MNm^{-2}	10^6 lb in^{-2}
Glass-ceramics	8·3–13·8	12–20
Fused silica glass	7·2	10·5
Soda-lime-silica glass	6·9	10·0
Borosilicate glass (Pyrex)	6·6	9·5
Sintered pure aluminia (5% porosity)	36·6	53
High alumina ceramics	27·6–36·6	40–50
Sintered pure magnesia (5% porosity)	20·7	30
Sintered beryllia (5% porosity)	31·0	45
Electrical high tension porcelain	6·6	9·6
Steatite low loss ceramic	6·9	10

For glasses, the modulus of elasticity, E, shows a roughly additive relationship with chemical composition and factors have been derived which enable the modulus to be calculated from the glass composition. The modulus of elasticity of a polyphase ceramic will also be an additive function of the individual characteristics of the crystalline and glassy phases and this has been confirmed by Freiman and Hench (1972). In a glass-ceramic it is to be expected that the modulus of elasticity will be determined primarily by the elastic constants of the major crystalline phases although the presence in the glass phase of oxides which promote the development of high values of E must be allowed for; in particular, calcium oxide, magnesium oxide and aluminium oxide appear to exert a marked influence upon the elastic moduli of glasses.

Data given in Table XIX show that significant differences exist in the elastic properties of glass-ceramics depending on their constitution. Thus, glass-ceramic 9608 which contains a β-spodumene/silica solid solution as the major phase has a significantly lower elastic modulus than glass-ceramic 9606 where the major crystalline phases are cordierite, cristobalite and rutile. Other examples of the dependence of elastic modulus upon glass-ceramic type are given in Table XX. Without knowledge of the volume fractions of the various crystal and residual glass phases it is not possible to form a complete idea of the relationship between microstructure and elastic properties. A more useful approach is to consider the influence of systematic variations of the composition and microstructure.

Table XIV gives the results of one such study. A marked increase of the elastic modulus occurs when the ZnO content exceeds the Al_2O_3 content and it is likely that this results from the appearance of quartz as a major phase and the disappearance of β-spodumene.

Variation of the heat-treatment schedule of a glass-ceramic allows different volume fractions of crystal phases to be developed and therefore permits the influence upon elastic properties to be examined. McMillan *et al.* (1966c) investigated the effects of heat-treatment of a glass-ceramic of the molecular percentage composition: SiO_2 72·5, Li_2O 22·3, K_2O 1·5, ZnO 3, P_2O_5 0·7. Values of the Young's modulus, E, were determined for specimens subjected to the heat-treatments given in Table XXI. The marked increase of E resulting from heat-treatment at 600°C was attributed to the development of lithium disilicate crystals. At 700°C and upwards silica crystallises out giving quartz in the final material. The development of this phase appears to result finally in a small reduction of the elastic modulus.

Borom *et al.* (1975) gave data on the elastic moduli of glass-ceramics derived from a modified Li_2O–SiO_2 glass. The elastic modulus of a glass-ceramic containing 35 weight per cent lithium disilicate was $9{\cdot}0 \times 10^4$ MNm^{-2} (13×10^6 lb in^{-2}) as compared with $7{\cdot}0 \times 10^4$ MNm^{-2} ($10{\cdot}2 \times 10^6$ lb in^{-2}) for the parent glass. A glass-ceramic containing 29 weight per cent $Li_2O.2SiO_2$ and 2 weight per cent $Li_2O.SiO_2$ gave a value of $8{\cdot}6 \times 10^4$ MNm^{-2} ($12{\cdot}5 \times 10^6$ lb in^{-2}) while one containing 20 weight per cent $Li_2O.SiO_2$ and no disilicate gave a value of $7{\cdot}9 \times 10^4$ MNm^{-2} ($11{\cdot}4 \times 10^6$ lb in^{-2}).

Using a glass of the molecular percentage composition SiO_2 61·0, Li_2O

TABLE XIX

ELASTIC CONSTANTS FOR GLASS-CERAMICS MEASURED AT 20°C

Constant	Glass-ceramic 9606	Glass-ceramic 9608
Modulus of elasticity, E	$12{\cdot}0 \times 10^4$ MNm^{-2} ($17{\cdot}4 \times 10^6$ lb in^{-2})	$8{\cdot}6 \times 10^4$ MNm^{-2} ($12{\cdot}5 \times 10^6$ lb in^{-2})
Modulus of rigidity, G	$4{\cdot}8 \times 10^4$ MNm^{-2} ($6{\cdot}9 \times 10^6$ lb in^{-2})	—
Bulk modulus, K	$7{\cdot}9 \times 10^4$ MNm^{-2} ($11{\cdot}4 \times 10^6$ lb in^{-2})	—
Poisson's ratio	0·245	0·25

TABLE XX

YOUNG'S MODULUS, E FOR VARIOUS GLASS-CERAMICS MEASURED AT 20°C

Type of glass-ceramic	Young's modulus, E		Major crystal phases present in glass-ceramic
	MNm^{-2}	$lb\ in^{-2}$	
$Li_2O–Al_2O_3–SiO_2$	$10{\cdot}4 \times 10^4$	$15{\cdot}1 \times 10^6$	β-Spodumene, lithium metasilicate
$Li_2O–MgO–SiO_2$	$9{\cdot}5 \times 10^4$	$13{\cdot}8 \times 10^6$	Quartz, lithium disilicate
$Li_2O–ZnO–SiO_2$:			
(*i*) Low ZnO content	$8{\cdot}7 \times 10^4$	$12{\cdot}6 \times 10^6$	Lithium, disilicate, cristobalite, quartz
(*ii*) High ZnO content	$7{\cdot}3 \times 10^4$	$10{\cdot}6 \times 10^6$	Cristobalite, lithium zinc silicate, (Li_2O, ZnO, SiO_2)
$Li_2O–ZnO–Al_2O_3–SiO_2$	$8{\cdot}3 \times 10^4$	$12{\cdot}0 \times 10^6$	Silica-rich β-spodumene solid solution; lithium disilicate

TABLE XXI

CHANGE IN ELASTIC MODULUS WITH HEAT-TREATMENT

	E:MNm^{-2}	lb in^{-2}
Chilled glass "as drawn"	$7{\cdot}2 \times 10^4$	$10{\cdot}4 \times 10^6$
Annealed at 450°C for 1 hr	$7{\cdot}6 \times 10^4$	$11{\cdot}0 \times 10^6$
500°C, 1 hr; 600°C, 1 hr	$9{\cdot}2 \times 10^4$	$13{\cdot}3 \times 10^6$
500°C, 1 hr; 700°C, 1 hr	$9{\cdot}6 \times 10^4$	$13{\cdot}9 \times 10^6$
500°C, 1 hr; 700°C, 2 hr	$9{\cdot}8 \times 10^4$	$14{\cdot}2 \times 10^6$
500°C, 1 hr; 800°C, 1 hr	$9{\cdot}0 \times 10^4$	$13{\cdot}0 \times 10^6$

30·5, K_2O 1·5, Al_2O_3 1·0, P_2O_5 1.0, B_2O_3 5.0 Atkinson and McMillan (1976) investigated the effects of heat-treatment in the range 600–900°C on the elastic modulus. The results summarised in Fig. 61 show that the modulus rises to a maximum for a heat-treatment temperature of 750°C. In this temperature range, the major crystal phase formed was lithium disilicate and microstructural studies showed that the volume fraction of crystalline phase in the glass-ceramic was a maximum after heat-treatment at 750°C. The variation of elastic modulus was, however, somewhat larger than would be expected simply from the observed variation of crystal volume fraction alone. Also the increase in elastic modulus when the heat-treatment temperature was

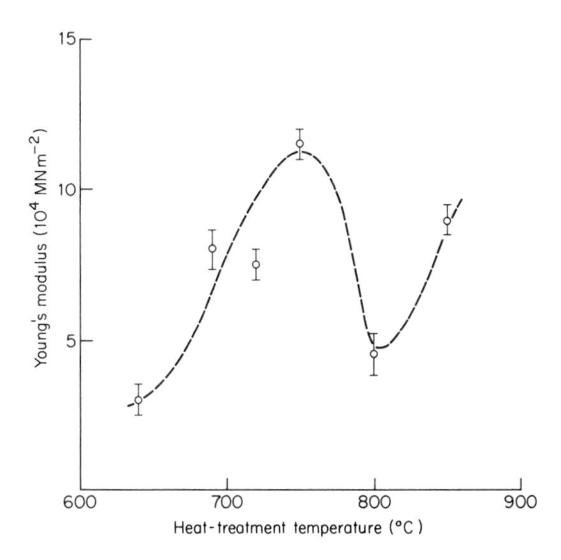

FIG. 61. Elastic modulus as a function of heat-treatment temperature for a glass-ceramic.

increased above about 820°C is difficult to explain purely in terms of the variation of the volume fraction of crystalline phase present.

c. The influence of temperature upon elastic properties. The effect of temperature upon the elastic constants of glass-ceramics can in some cases reveal marked influences resulting from the presence of certain crystal phases.

The curves in Fig. 62 are for different types of phosphate-catalysed glass-ceramics. Perhaps the most significant feature of these data is the presence of marked minima in the elastic moduli of compositions A and B at a temperature in the region of 200°C, while compositions C and D do not show this minimum. These contrasting effects are related to the crystal contents of the different glass-ceramics and it is significant that while compositions A and B contain cristobalite as a major phase, this crystal phase is not present in compositions C and D. It is very probable that the minimum value of the elastic modulus which occurs at about 200°C is associated with the α-β cristobalite inversion which can occur at temperatures within the range 220°C to 270°C depending upon the purity of the cristobalite. No doubt other

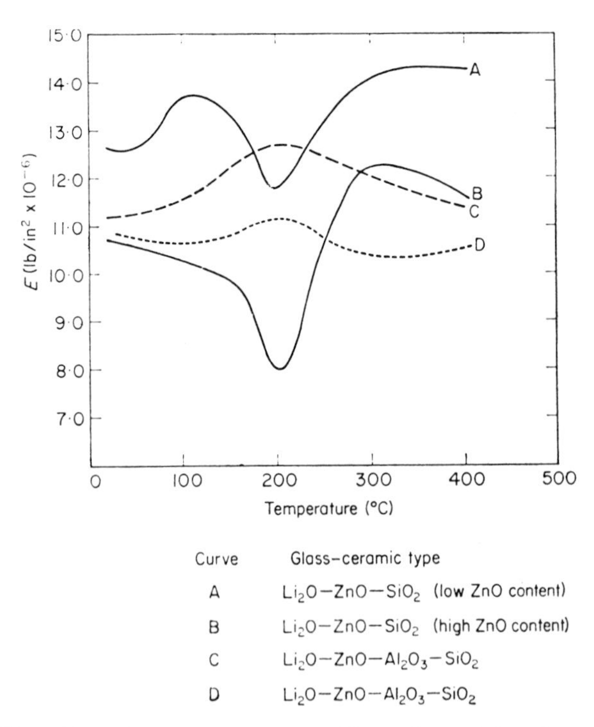

FIG. 62. The effect of temperature upon Young's modulus for various glass-ceramics.

crystals which exhibit phase inversions could also lead to minimum values of the elastic modulus of the glass-ceramic. It is known, for example, that crystalline quartz shows a marked minimum in the value of the modulus of elasticity at a temperature of about 570°C according to data given by Vigoreaux and Booth (1950). The α-β quartz inversion occurs at a temperature of 573°C, which coincides quite closely with the temperature at which the minimum modulus of elasticity occurs. Thus in a glass-ceramic containing quartz as a major phase, it is possible that a minimum in the elastic modulus-temperature curve would occur at about 570°C.

The data given in Fig. 63 for glass-ceramic 9606, a magnesium-aluminosilicate type glass-ceramic, show that for this material also there is a marked minimum in the Young's modulus/temperature curve at a temperature of about 160°C. A minimum value of the shear modulus occurs at a similar temperature and the value of Poisson's ratio shows a sudden increase at a temperature of 160 to 170°C. These effects are probably related to a phase inversion occurring in the glass-ceramic, although the temperature at which

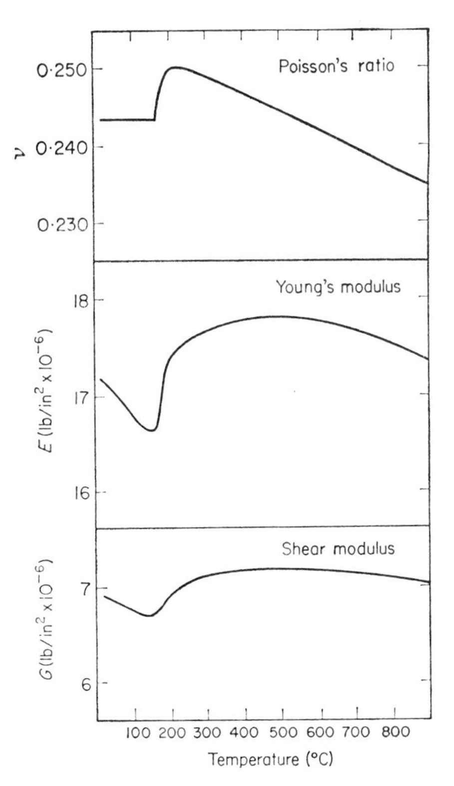

FIG. 63. The effect of temperature upon the elastic constants of glass-ceramic 9606.

the effects are observed is somewhat lower than that at which the cristobalite inversion is usually noted.

Muchow (1971) has reported the variation of the elastic moduli as a function of temperature in the range of 20 to 800°C for glass-ceramics containing keatite solid solutions of the general formula $Li_2O.Al_2O_3.nSiO_2$ where n ranged from 4 to 8. The material for which n equalled 4 (beta-spodume composition) showed a gradual fall of elastic modulus from $9{\cdot}2 \times 10^4$ MNm^{-2} ($13{\cdot}4 \times 10^6$ lb in^{-2}) at 20°C to $9{\cdot}1 \times 10^4$ MNm^{-2} ($13{\cdot}2 \times 10^6$ lb in^{-2}) at 800°C. The higher members of the series showed anomalous behaviour in that the elastic moduli increased with increase of temperature, the effect being most marked for the material for which n equalled 10. In this case the value increased from $7{\cdot}6 \times 10^4$ MNm^{-2} (11×10^6 lb in^{-2}) at 20°C to $8{\cdot}3 \times 10^4$ MNm^{-2} (12×10^6 lb in^{-2}) at 800°C. The materials also exhibited anomalous isothermal compressibility effects in addition to unusual thermal expansion behaviour; they have low or negative thermal expansion coefficients. Muchow considers that the anomalous thermal and elastic properties of the glass-ceramics containing keatite solid solutions are due to changes in the Si–O–Si bond lengths and bond angles.

3. *Hardness and Abrasion Resistance*

a. General considerations. Hardness is not a fundamental physical characteristic of a material but rather it is a complex function of a number of physical properties which are combined in differing degrees depending on the method of test. In a sense, we can equate the hardness of a material with its resistance to abrasion or wear and this characteristic is of practical interest since it may determine the durability of a material during normal use and it may also decide the suitability of the material for special applications where abrasion resistance is of prime importance. Thus, the resistance of the material to abrasion under closely specified conditions may be taken as one measure of hardness. The resistance to scratching of the material by other materials represents another measure of its hardness, and the depth of penetration of a loaded pyramid (usually a diamond) into the material under specified conditions represents yet another measure of hardness. All of these methods are empirical and, although any one of them can be used to classify materials in a scale of hardness, the order of hardness may differ according to the method employed. For this reason care must be taken to select test methods which represent the closest approximation to the practical conditions of use wherever possible.

b. Hardness tests. Scratch hardness has the merit of being relatively simple to carry out but the results are often of uncertain value. The Moh scale of hardness, in which minerals are classified according to their ability to scratch other minerals, is well known. In this scale, diamond at the upper end has a hardness of 10 and talc at the lower end has a hardness of 1. Glasses usually have Moh's hardness numbers of 5 to 7, but the test is not particularly suitable for classifying individual glasses with respect to one another.

Glass-ceramics tested in this fashion appear to have scratch hardnesses which are considerably higher than those of ordinary glasses. Thus Watanabe *et al.* (1962) found that a lithium-aluminosilicate glass-ceramic resisted scratching by materials with hardness less than 9 on Moh's scale, so that in this respect the glass-ceramic was comparable with corundum (Al_2O_3) with a hardness of 9 or silicon carbide with a hardness of 9 to 9·5. Earlier, Stookey (1956) had stated that for photosensitively nucleated glass-ceramics, the hardness on Moh's scale was about 7·5 as compared with 5·5 for the original glasses.

Attempts have been made to devise more reliable scratch hardness tests than that proposed by Moh; such methods are based on studying the effects of drawing a diamond scribe across the glass under specified conditions. Unfortunately, no tests of this type on glass-ceramics have so far been reported.

Indentation hardness has been widely applied in the metallurgical field and this method has also been applied to glasses and similar materials. Under suitable conditions, the point of a diamond pyramid will penetrate a glass surface leaving a permanent indentation. The mechanism by which the indentation is formed is complex but it combines the effects of elastic deformation, plastic flow and compaction of the glass beneath the indenter.

A number of indentation hardness test methods are available; both the Knoop hardness and the diamond pyramid hardness (D.P.H.) depend on measuring the size of the permanent indentation produced although differently shaped indenters are used. The Knoop hardness is expressed as the load in kilogrammes divided by the projected area of indentation in square millimetres, whereas the D.P.H. is expressed as the load divided by the contact area. In the Rockwell test a spheroconical diamond indenter under a load of 150 kg is employed. The depth of penetration of the indenter into the test-piece under specified conditions serves to define the hardness. None of these methods can be regarded as truly satisfactory for measuring the hardness of non-ductile materials such as glass-ceramics. However, measurements of this type have been made and it is worthwhile to note them, bearing in mind their limitations.

Table XXII gives value for the indentation hardnesses of glasses and glass-ceramics. The hardness values of the two glass-ceramics are significantly greater than those of the two glasses although they are lower than that of the high alumina ceramic. It should be noted that the hardness value is dependent

TABLE XXII

INDENTATION HARDNESS OF GLASSES AND GLASS-CERAMICS

Material	Test load (g)	Hardness Knoop (kg/mm^2)	D.P.H. (kg/mm^2)
Glass-ceramic 9606	100	689	631
	500	619	—
Glass-ceramic 9608	100	703	707
	500	588	639
Glass 7900	100	532	—
(96% silica glass)	500	477	—
Glass 7740	100	481	—
(low expansion borosilicate glass	500	442	—
High alumina ceramics,	100	1880	—
(>93% AL_2O_3)	500	1530	—

to some extent upon the load employed during the test, so that when the indentation hardnesses of materials are compared the test conditions must be specified. Glass-ceramics are apparently "harder" than gray cast-irons, for which the Knoop hardness (500 g load) ranges from 180 to 300 kg/mm^2, or annealed stainless steel for which the corresponding Knoop hardness is 150–200 kg/mm^2.

A study of the variation of indentation hardness as a function of heat-treatment has been carried out by Atkinson and McMillan (1976) for a glass-ceramic based on the Li_2O–SiO_2 system. A small but significant increase of the Vickers Hardness occurred as the heat-treatment was increased from 600° to 700°C. Further increases of the heat-treatment temperature up to a maximum of 900°C resulted in a reduction of hardness.

The maximum hardness corresponded approximately but not exactly with the maximum volume fraction of the crystal phase in the glass-ceramic suggesting that microstructural effects influence hardness. Other work in the author's laboratory has indicated that the nature of the crystal phases may play a highly significant role in determining hardness. Some crystal phases, even when present in only a small volume fraction seem to result in marked enhancement of hardness. Phases of the spinel type appear to be particularly effective in this respect.

The Rockwell C hardness test has also been used to compare various glass-ceramics (Partridge, 1961). It was found that the Rockwell C hardnesses for glass-ceramics ranged from 60 upwards. For comparison purposes the Rockwell C hardness value for air-hardened gear steel is 50. Caution should be exercised in making comparisons of this type, however, in view of the very different natures of glass-ceramics and metals. There appeared to be differences in the "hardnesses" of different glass-ceramics as measured by the Rockwell C test. A lithium-aluminosilicate material gave a significantly higher value than a lithium-zinc-silicate composition and this in turn was apparently "harder" than a lithium-magnesium-silicate composition.

c. Abrasion and resistance to wear. It is thought that a more practical and reliable method of studying the relative hardnesses of glass-ceramics would be to abrade them under standard conditions and to measure the amount of wear which occurs. Using a specially designed apparatus in which a disc of the material is rotated beneath a loaded tungsten carbide abrading head such comparisons have been made. These tests have shown that glass-ceramics have good abrasion resistance and in a test in which the specimen was rotated for 50 000 revolutions followed by measurement of the volume of material removed by abrasion, it was found that a lithium-zinc-silicate glass-ceramic had an abrasion resistance ten times that of 18–8 stainless steel. A lithium-aluminosilicate glass-ceramic had an even better abrasion resistance and there appeared to be some correlation between the abrasion resistance measured in this way and the Rockwell C hardness measurements described previously.

In considering the possibility of wear occurring between glass-ceramic parts which are moving in contact with one another due to sliding or rotation, the extent to which friction occurs becomes important. For this reason, it is of interest to know the values for coefficients of friction between glass-ceramic surfaces. Table XXIII gives values for the static and dynamic coefficients of

TABLE XXIII

COEFFICIENT OF FRICTION FOR GLASS-CERAMICS

Glass-ceramic type	Coefficient of friction	
	Static	Dynamic
Lithium-zinc-silicate (low ZnO content)	0·19	0·16
Lithium-zinc-silicate (high ZnO content)	0·09	0·07
Lithium-aluminosilicate	0·16	0·14

friction measured between polished surfaces of different types of glass-ceramic.

The differences observed for the coefficient of friction values may be partly determined by the relative ease with which the glass-ceramics can be polished to have smooth surfaces.

C. Electrical Properties

1. *Electrical Resistivity*

a. General considerations. If a glass-ceramic is to be used for electrical insulation, its resistivity should be as high as possible. In many cases the insulating material is required to operate at or near to normal ambient temperatures but in some devices the insulator may be required to operate at elevated temperatures, and for these applications the variation of electrical resistivity with temperature is important. Conduction may take place across the surface or through the volume of the material so that, in assessing the electrical insulating qualities, the surface resistivity and the volume resistivity must be taken into account. The surface is defined as the resistance in ohms of a strip of the surface of unit length and width and the units are ohms per square. The volume resistivity is the longitudinal resistance in ohms per unit length of a uniform bar of unit sectional area; the unit of volume resistivity is the ohm cm.

b. Surface resistivity. The characteristic can be important since it is often observed for materials which are used for insulators that the leakage of current across the surface is greater than that passing through the body of the material. This effect is particularly noticeable at room temperature in humid atmospheres and the surface conductivity is largely caused by the adsorption of moisture and other substances from the atmosphere. If leaching of the surface of the insulator occurs, so that alkali metal ions become dissolved in the thin water layer, the surface conductivity is greatly increased; this effect is particularly marked for soda-lime-silica glass and for other glasses with relatively high alkali contents. Similarly, for glass-ceramics, it would be expected that the surface resistivity would be influenced by the possibility of surface leaching and the production of a water layer containing alkali metal ions. Obviously the conditions of test will be important since the surface resistivity measured under completely dry conditions will be much higher than that measured under humid conditions. Few measurements of surface resistivity for glass-ceramics have been reported, but the values given in Table XXIV show that a lithium-zinc-silicate composition has a high surface

TABLE XXIV

SURFACE RESISTIVITIES OF VARIOUS MATERIALS MEASURED AT 20°C AND 70 PER CENT RELATIVE HUMIDITY

Material	Surface resistivity (ohms per square)
Glass-ceramic	10^{16}
Borosilicate electrical glass	$10^{11.7}$
Soda-lime-silica-glass	$10^{9.7}$
Glazed electrical porcelain	10^{14}

resistivity under conditions of 70 per cent humidity and that it is superior in this respect to a glazed electrical porcelain and to a borosilicate electrical glass. Of course, not all glass-ceramics will show such high values of surface resistivity, particularly if the residual glass phase in the material contains appreciable proportions of alkali metal oxides such as Na_2O or K_2O, but, in general, glass-ceramics will have adequate surface resistivities for electrical insulation purposes.

Although surface resistivity is chiefly important when the material is to be used at normal ambient temperatures, there are cases where low surface electrical leakage is necessary at high temperatures. In this connection, it has been found that the surface resistivity of a lithium-zinc-silicate glass-ceramic remains quite high even at temperatures up to 700°C, as the data given in Table XXV show.

c. Volume resistivity. In general, glass-ceramics are good electrical insulators in comparison with glasses and conventional ceramics. This is illustrated by the data given in Table XXVI which shows that at temperatures up to 400°C

TABLE XXV

EFFECT OF TEMPERATURE ON THE SURFACE RESISTIVITY OF A LITHIUM-ZINC-SILICATE GLASS-CERAMIC

Temperature (°C)	Surface resistivity (ohms per square)
100°C	$10^{15.4}$
300°C	$10^{10.2}$
500°C	$10^{8.1}$
700°C	$10^{7.3}$

TABLE XXVI

ELECTRICAL RESISTIVITIES OF GLASS-CERAMICS AND OTHER MATERIALS

Material	Log$_{10}$ resisitivity at: 20°C	200°C	400°C	600°C	800°C
Glass-ceramics					
(*a*) Lithium-alumino-silicate type	13·7	8·5	5·3	4·1	—
(*b*) Lithium-zinc-silicate type; low ZnO content (c. 5%)	15·5	11·5	7·7	5·0	3·7
(*c*) Lithium-zinc-silicate type: high ZnO content (c. 30%)	13·8	—	6·7	4·7	4·2
(*d*) Lithium-zinc-lead silicate	—	10·9	8·5	—	—
Glasses					
(*a*) Soda-lime-silica glass	—	6·9	4·2	2·7	—
(*b*) Alkali-lead-silicate glass	—	11·0	6·9	3·9	2·5
(*c*) Low loss electrical borosilicate glass	—	9·8	6·8	4·4	—
Ceramics					
(*a*) High alumina ceramic (95% Al_2O_3)	15·3	11·9	9·3	7·8	6·9
(*b*) High tension porcelain	13·7	8·6	5·5	4·2	—

or so, the glass-ceramics have resistivities that are similar to those of other good insulating materials.

The volume conductivities of oxide glasses and ceramics generally result from the transport of mobile ions through the materials. There are, however, glasses, ceramics and glass-ceramics in which the transport process involves transport of electrons or positive holes and which are therefore classed as semiconductors. The great majority of glass-ceramics are ionic conductors and it is this process that will mainly concern us.

For glasses and ceramics, it is found that the mobile alkali metal cations (eg Li^+, Na^+) contribute strongly to the conduction process and the conductivity of glass increases rapidly as the alkali metal oxide content is increased. It is likely for glass-ceramics also that the alkali metal cations exert a dominant effect on conductivity.

The mobility of the ions is strongly dependent on the structure through which they are migrating. In a crystal, an ion leaving an interstitial site has to pass through an intermediate position of high energy in order to enter an adjacent site of lower energy. At a given temperature a proportion of the ions possess sufficient thermal energy to surmount this energy barrier and can move into vacant sites. Random wandering of the mobile ions therefore occurs, but when a dc field is applied a net migration in the direction of the field gradient takes place and therefore conduction occurs.

A similar process takes place for glasses but with the difference that the energy barrier does not have a unique value owing to the random nature of the glass structure. Also, in glasses the average energy barrier is lower than in crystals giving the alkali metal ions greater mobility and resulting in higher conductivity than for a crystal of the same composition.

Of the important factors that determine conduction, mobility is therefore certain to be affected when the glass crystallises. Generally we expect this to be reduced and therefore electrical resistivity to increase when the glass structure is converted into crystalline phases.

Another factor likely to influence the conductivity of a glass-ceramic is the presence of phase boundaries. These may play a part in hindering motion of the ions, especially if boundary stresses are present, and thus reduce the conductivity still further.

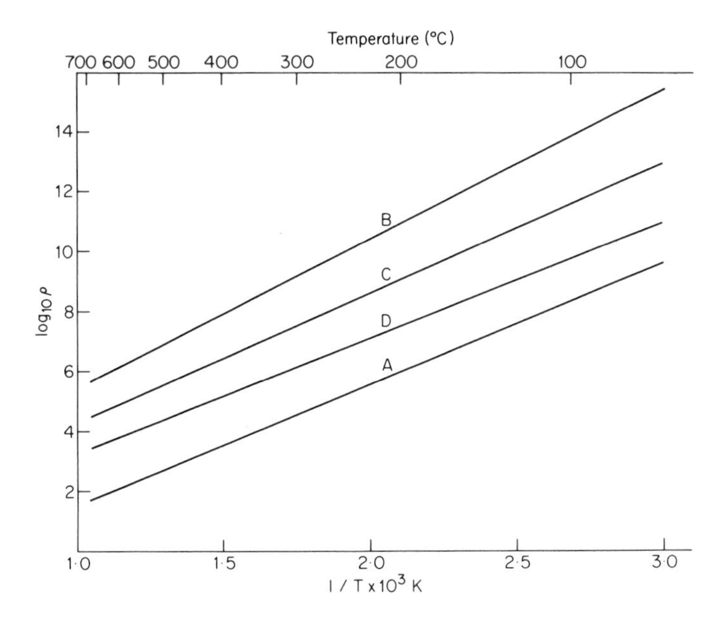

FIG. 64. Resistivity-temperature relationships for glasses and glass-ceramics: A and C are for the parent glasses; B and D are for the corresponding glass-ceramics.

For both glasses and crystals the ionic conductivity increases rapidly as the temperature is raised and the temperature dependence of conductivity is given by

$$\sigma = \sigma_0 \exp(-E/KT)$$

where σ_0 is a term not strongly dependent on temperature and E is the activation energy of conduction. Thus a plot of log σ (or of log ρ, the resistivity) versus the reciprocal of the absolute temperature will give a straight line the slope of which gives E.

Curves of this type confirm the normally observed reduction of electrical resistivity on conversion of a glass to a glass-ceramic (Fig. 64) curves A and B. The volume resistivity of the glass-ceramic is at least four orders of magnitude greater than that of the parent glass in the temperature range 20 to 400°C. The activation energy of conduction for the glass-ceramic (23·3 Kcal mole^{-1}) is also greater than that for the glass (19·3 Kcal mole^{-1}).

The increase of activation energy E, would not be sufficient to account for the very large increase of resistivity. There is also a change in the pre-exponential term σ_0 in the expression relating resistivity, σ, with temperature. This term can be expressed as:

$$\sigma_0 = e^2\lambda^2 n\nu/2KT$$

in which λ is the ionic jump distance, n is the number of atoms per unit volume and ν is the vibrational frequency of the ion in its well. It seems likely that depletion of the intercrystalline glass phase of mobile lithium ions (that is reduction of n) is responsible for the increase of the pre-exponential term. In this glass-ceramic the lithium ions are largely incorporated in lithium silicate crystals where they are thought to be tightly bound and therefore to make only a small contribution to the conductivity.

Similar large increases of electrical resistivity are observed for other glass-ceramics where the alkali ions are incorporated into crystal phases. However, in cases where these ions do not take part in a crystal phase, the effect of crystallisation is to increase the concentration of the alkali metal ions in the residual glass phase. In such cases, the resistivity of the glass-ceramic is lower than that of the parent glass. Curves C for a glass and D for the corresponding glass-ceramic in Fig. 64 illustrate this effect. The glass-ceramic in question is of the ZnO–Al_2O_2–SiO_2 type containing a small proportion of sodium oxide. In this case the reduction of the resistivity of the material on conversion to the glass-ceramic is largely caused by the reduction in the activation energy of conduction from 17·2 to 15·4 Kcal mole^{-1} because the values of σ_0 the pre-exponential constant are virtually identical. This type of behaviour is comparatively rare; the more general effect is for the glass-ceramic to have a very much higher resistivity than the parent glass.

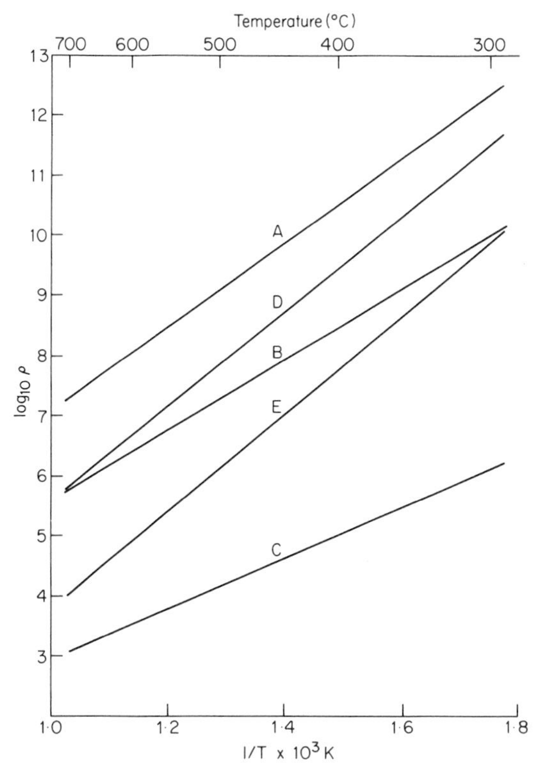

FIG. 65. Resistivity-temperature relationships for glass-ceramics.

Depending on the chemical composition and on the nature of the crystal phases present, the electrical resistivities of glass-ceramics can vary widely. Values of resistivity at 300°C, for example, range from 10^6 to 10^{12} ohm cm. Figure 65 gives some examples of resistivity versus temperature relationships for a range of glass-ceramics. Curve A is for an alkali-free zinc-alumino-silicate glass-ceramic in which the major crystal phase is willemite together with some gahnite. The resistivity of this material at 300°C (2×10^{12} ohm cm) is higher than that of vitreous silica (10^{11} ohm cm) and thus it is a very good electrical insulator. Alkali-free glass-ceramics of the cordierite type possess similar high electrical resistivities. If small amounts of sodium oxide are introduced into the ZnO–Al_2O_3–SiO_2 glass-ceramic, marked reductions of resistivity occur as shown by curves B and C in Fig. 65 which are for glass-ceramics containing 1 and 2 weight per cent Na_2O respectively. In this case, the sodium ions are thought to concentrate in the residual glass phase and this accounts for the remarkably large effects of the small additions.

Glass-ceramics with high lithium contents tend to have relatively low electrical resistivities as might be expected. Curve D in Fig. 65 is for a low expansion glass-ceramic containing a β-spodumene/silica solid solution as the major phase and the value of the resistivity at 300°C in this case is 10^5 ohm cm. Work by Johnson *et al.* (1975) on a glass-ceramic containing β-eucryptite as the major crystal phase showed that this material had a relatively high electrical conductivity. At 300°C the resistivity of the material was $\sim 10^5$ ohm cm. These investigations suggest that the contribution to the conductivity from the glass phase was greater than that from the crystalline phase and also that grain boundary effects could be important. Studies by Atkinson and McMillan (1976) on a glass-ceramic containing 30 mol. per cent lithia showed that the electrical resistivity was markedly dependent on the heat-treatment process used. Crystallisation heat treatments in the range 600 to 900°C were employed and it was found that the electrical resistivity measured at 300°C attained a clearly defined maximum of $\sim 10^6$ ohm cm for specimens heat-treated at 750°C. This was accompanied by a maximum in the activation energy for conduction. Microstructural analysis showed that the volume fraction of crystalline phase also attained a maximum for heat-treatment temperatures in the same region.

As will have become clear from the preceding discussion the curve relating to log resistivity with the reciprocal of absolute temperature for the majority of glass-ceramics approximates very closely to a single straight line for a wide temperature range implying a single value of activation energy. Exceptions to this rule exist, however, and an example is given by curve C in Fig. 66. Here the data are best represented by two straight lines intersecting at a temperature of $\sim$ 240–250°C. This material, which was of the Li_2O–ZnO–SiO_2 type containing about 15 weight per cent ZnO, contained cristobalite as a major phase and the discontinuity in the resistivity-temperature is most probably related to the α-β cristobalite inversion which occurs at temperatures between 220 and 270°C.

Although emphasis has been placed on the need for a low alkali metal ion content in glass-ceramics in order to achieve low electrical conductivity, in some cases it may not be possible to avoid the inclusion of these ions. For example, their presence may be necessary to achieve a particular thermal expansion coefficient (required if the glass-ceramic is to be sealed to a metal). In such cases the electrical resistivity can be considerably increased by including in the parent glass large, relatively immobile ions such as those of lead or barium. These ions, which tend to become concentrated in the residual glass phase of the glass-ceramic, block the migration of the alkali metal ions and therefore result in a reduction of the ion conductivity. Curve E in Fig. 65 is for a glass-ceramic containing about 25 mole per cent Li_2O but since lead oxide is also present in the material, a fairly high resistivity of 3×10^9 ohm cm at 300°C is achieved.

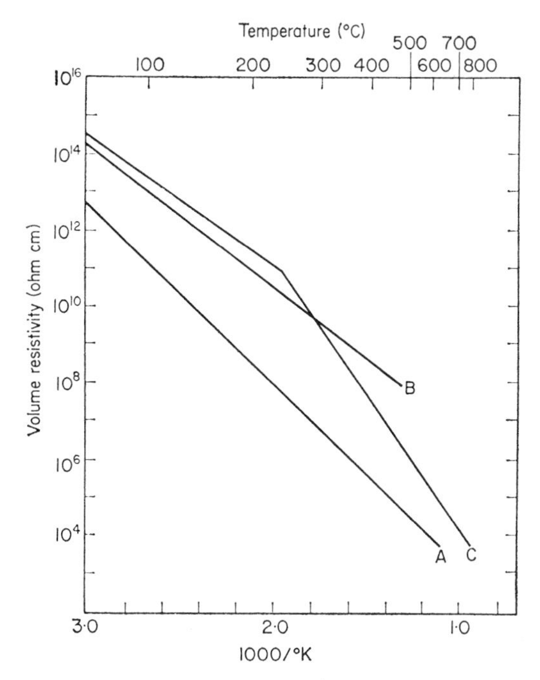

FIG. 66. Volume resistivity versus temperature for selected glass-ceramics.

In cases where controlled electrical conductivity is desired, the use of ionically conducting materials is undesirable because of polarisation effects resulting from depletion of the current carrying ions in regions close to the electrodes. Semiconducting glass-ceramics do not of course exhibit this phenomenon. Such materials contain at least one crystalline phase which is itself a semiconductor and this phase must be present in a sufficiently high concentration to form a continuous network throughout the bulk of the glass-ceramic.

One semiconducting glass-ceramic is prepared from a calcium borosilicate glass containing ~ 12 per cent of ferric oxide. Heat-treatment of this composition at 800°C results in a large reduction of the volume resistivity at room temperature to a value of 10^6 ohm cm. The high electrical conductivity is attributed to the formation of an interconnected network of Fe_3O_4 crystals. Other semiconducting glass-ceramics are known but so far these do not seem to have achieved widespread application.

In some cases, the presence of semiconducting crystals in the glass-ceramic may occur unintentionally and can be highly undesirable giving rise to unacceptably high dielectric losses.

2. *Dielectric Properties*

a. Dielectric loss. When an electric field is established in a dielectric, electrical energy is stored within the material and on removal of the field the energy may be wholly recoverable, but usually only part of the energy is recoverable; the part of the electrical energy which is lost appears as heat. In an alternating field, therefore, a power loss occurs for insulating materials. The ratio between the irrecoverable and recoverable parts of the electrical energy is expressed as $\tan \delta$, the power factor. The dielectric losses are also dependent upon K, the permittivity or dielectric constant of the material and the dielectric loss factor is given by the product of the permittivity and power factor, $K \tan \delta$. Low dielectric losses are essential if an insulating material is to be used for high-frequency insulation and for this purpose it is desirable that the values of both $\tan \delta$ and K should not increase markedly with increase of temperature. This is necessary, of course, in cases where the dielectric may be required to operate at high temperatures but it is also necessary when the insulator is to be used at normal temperatures since some losses, causing heating of the dielectric, will inevitably occur.

There are three sources of energy loss in a dielectric such as glass:

(*i*) Ion migration losses.

(*ii*) Resonance type ion vibration losses.

(*iii*) Deformation losses related to the deformation of the network or lattice.

Of these, the ion migration losses are by far the most important in the frequency range which is concerned in electronics. The ion vibration losses and the deformation losses are not important for frequencies below 10^{10} cycles per second. For glasses it is found that a correlation exists between the power factor, $\tan \delta$, and the electrical resistivity and this supports the idea that the dielectric losses are related to ion migration processes. In glasses, therefore, it is the more mobile ions which are largely responsible for the dielectric losses in the intermediate frequency range. These ions include the alkali metal ions and the dielectric losses increase in the order of mobility of the cations which is:

$$Rb^+ \rightarrow K^+ \rightarrow Na^+ \rightarrow Li^+$$

The mobility of these ions is also influenced by the network structure of the glass and changes which lead to a more open and less strongly bonded network tend to cause increased dielectric losses. The dielectric losses are also affected by the presence of other modifier cations which occupy interstitial positions and large ions, such as those of barium and lead, are particularly effective in reducing the losses since they tend to block the motion of the mobile alkali metal ions. For a glass-ceramic, the position is more complicated than for a glass, and it is difficult to assess the exact effects of the various crystal phases and of the residual glass phase, the precise chemical composition of which is not usually known. The dielectric losses for single crystals are small and are mainly determined by minor impurities, and it is likely that the glass phase of the glass-ceramic is the main contributor to dielectric losses. This means that to achieve low losses the amount of residual glass phase must be as small as possible and its composition must be controlled to avoid, as far as is practicable, the presence of major proportions of alkali metal ions. In addition to the effects of chemical constitution, the dielectric losses of a glass-ceramic, like those of a glass, are related to the frequency at which they are measured and to the temperature. The general effect is for the losses to increase as the temperature is raised, although the temperature dependence of dielectric losses becomes less marked as the frequency increases.

The marked reduction of loss tangent that can accompany crystallisation of a glass, in cases where the alkali metal ions are incorporated into a crystal phase, has already been mentioned in chapter 3. This effect is further illustrated by the data given in Table XXVII for a glass and the glass-ceramic derived from it. The material is of the lithium-alumino-silicate type in which lithium-disilicate is present as a major phase. Clearly the incorporation of lithium ions in this crystal renders their mobility significantly less so that their contribution to dielectric loss is reduced.

In some cases where mobile alkali metal ions (eg Na^+) are not incorporated in a crystal phase, but become concentrated in the residual glass phase, glass-ceramics exhibiting high dielectric losses can result. This is illustrated by the results given in Fig. 67. Curve A is for an alkali free $ZnO–Al_2O_3–SiO_2$ glass-ceramic and it is seen that the value of loss tangent is low and is not

TABLE XXVII

POWER FACTOR, TAN δ, OF A GLASS-CERAMIC AND THE PARENT GLASS MEASURED AT 1 MHz

Temperature (°C)	Power factor, tan δ	
	Parent glass	Glass-ceramic
25	0·0400	0·0018
50	0·0562	0·0039
70	0·0835	0·0071
100	0·1450	0·0087
120	0·2050	0·0101
140	0·4250	0·0147

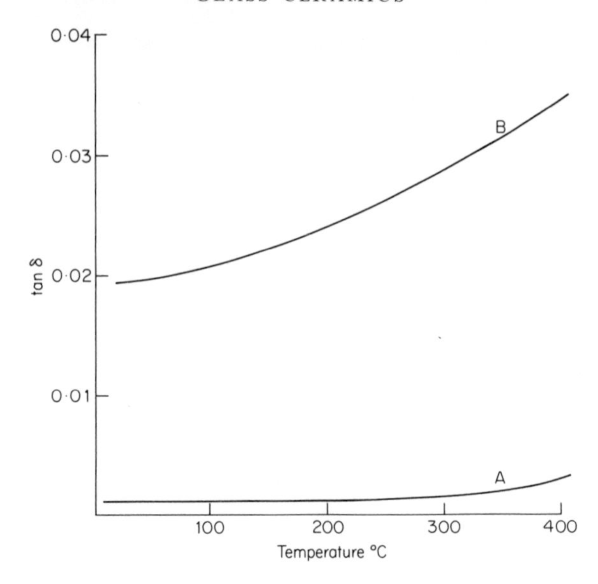

FIG. 67. Loss tangent-temperature relationship for glass-ceramics: A, alkali-free material, B, containing a small concentration of Na_2O (measurement frequency 10^4 MHz).

greatly affected by increase of temperature. Curve B, is for a glass-ceramic of the same general composition and contains the same crystal phases (willemite and gahnite) but it contains 2 per cent Na_2O. The effect of this is to increase the loss tangent almost by two orders of magnitude and also to result in a very marked rise of loss tangent with increasing temperature.

Table XXVIII gives data for a range of glass-ceramic types and other materials and shows that at the measurement frequency employed glass-ceramics exhibiting loss factors comparing favourably with those for low loss ceramics are available. The glass-ceramics contain approximately 10 weight per cent of Li_2O and it is apparent that the other modifying oxides present exert a significant influence upon the dielectric losses. In particular the incorporation of lead oxide causes a marked reduction of the dielectric loss and this can be attributed to the effect of lead ions in blocking oscillatory motions of lithium ions in the residual glass phase.

The dielectric losses of glass-ceramics containing alkali metal ions show a marked dependence upon frequency as shown by the curves in Fig. 68, which are for materials containing approximately 10 weight per cent of Li_2O. All of the materials show relatively high losses at low frequencies attributable to lithium ion motion. It is also evident that the dielectric losses increase markedly at frequencies above 10^9 Hz. While all of the materials show minimum dielectric losses at frequencies of 10^6 to 10^8 Hz, the composition

TABLE XXVIII

DIELECTRIC PROPERTIES OF GLASS-CERAMICS AND OTHER MATERIALS AT 1 MHz AND 20°C

Material	Power factor tan δ	Permittivity K	Loss factor K tan δ
Glass-ceramics			
Li_2O–Al_2O_3–SiO_2	0·0018	6·6	0·0119
Li_2O–MgO–SiO_2	0·0022	5·4	0·0119
Li_2O–ZnO–SiO_2 (c. 5% ZnO)	0·0023	5·0	0·0115
Li_2O–ZnO–SiO_2 (c. 10% ZnO)	0·0013	6·0	0·0078
Li_2O–ZnO–SiO_2 (c. 30% ZnO)	0·0063	5·3	0·0334
Li_2O–ZnO–PbO–SiO_2	0·0003	5·8	0·0017
Glasses			
Fused silica	0·0002	3·78	0·0008
Borosilicate low loss	0·0006	4·0	0·0024
Soda-lime-silica	0·0090	7·2	0·0648
Ceramics			
High aluminia (c. 95% Al_2O_3)	0·0004	8·8	0·0035
Steatite type	0·0013	5·9	0·0077
Forsterite type	0·0003	6·3	0·0019

containing lead oxide (curve C) is notable for the very wide frequency range (10^4 to 10^9 Hz) over which it exhibits low losses. Again, this valuable characteristic is attributable to the presence of relatively immobile lead ions in the glass phase.

It is to be noted that the variation of loss factor with frequency arises chiefly from variations of tan δ since the permittivity of these materials appears to be only weakly dependent upon the frequency. Thus, for the material represented by curve B, the values of permittivity at frequencies of 10^4, 10^6 and 10^9 Hz were 5·2, 5·0 and 5·5 respectively.

McMillan and Partridge (1972) investigated the dielectric properties of ZnO–Al_2O_3–SiO_2 glass-ceramics in which partial replacements were made for ZnO by CaO, MgO or BaO. The materials containing BaO were of especial interest since the values of loss tangent were low, generally less than 20×10^{-4} over a frequency range of 10^5 to 10^{10} Hz. It was observed, however, that a peak in the loss tangent versus frequency curve was displaced progressively towards lower frequencies, as the barium oxide content was increased (from $\sim 10^{9 \cdot 2}$ Hz for the baria-free base glass-ceramic to $\sim 10^{8 \cdot 5}$ Hz for a composition containing 10·6 mol. per cent BaO). It was proposed that the residual glass phase in these materials made the major contribution to dielectric loss and that the effect observed for barium oxide additions was due to a displacement of a maximum in the ionic resonance losses to lower frequencies. This was thought to be caused by the effects of barium oxide additions on the structure of the glass network, since no significant changes of the nature of the crystal phases formed could be detected.

It is of some interest to know whether the glass-ceramic microstructure, especially with regard to the mean sizes and volume fractions of crystal phases, is important in determining the magnitude of dielectric losses. Tashiro (1966) reported measurements of loss tangent for photosensitively nucleated (gold) glasses essentially of the Li_2O–Al_2O_3–SiO_2 type. He showed that the

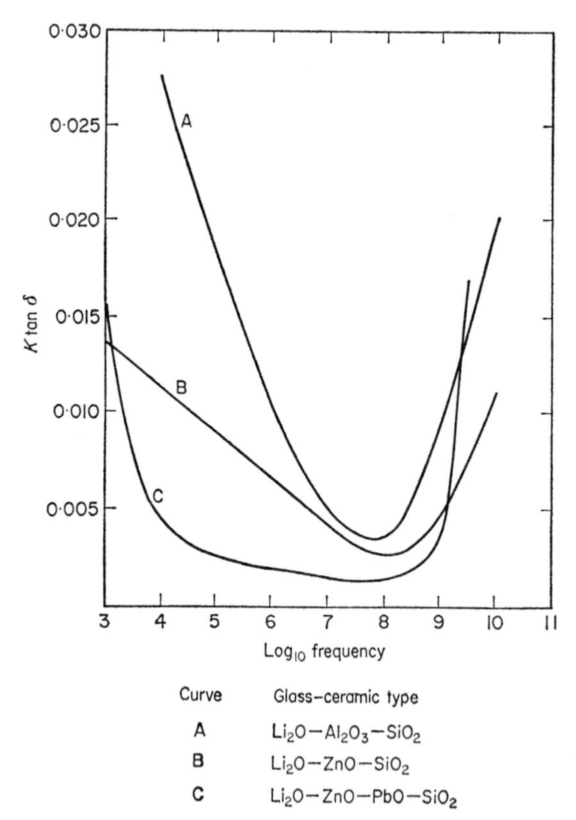

FIG. 68. Loss factor, $K \tan \delta$ versus frequency for various glass-ceramics.

value of tan δ was unaffected by a two-fold variation in crystal size, achieved by varying the time of exposure of the materials to ultraviolet radiation before heat-treatment. Since the glasses received identical heat-treatments it is presumed that the volume fraction of the crystal phase, lithium metasilicate for specimens heat-treated at 620°C, remained constant. A similar result was found for other glass-ceramics in which beta-spodumene and beta-eucryptite crystal phases were formed. It was however found that glass-ceramics containing a major proportion of the former crystal gave higher dielectric losses than those in which beta-spodumene predominated and it was suggested that in these materials the contributions of the crystal phases to dielectric loss were important. The value of the loss tangent at a frequency of 1 MHz for beta-eucryptite is greater than that for beta-spodumene by a factor of ten and this could account for the high dielectric losses of beta-eucryptite glass-ceramics.

The insensitivity of loss tangent to variations of the mean crystal size was confirmed by the work of Harper *et al.* (1970) who investigated the effects of variations in the heat-treatment cycle on the dielectric properties of P_2O_5-nucleated Li_2O–SiO_2 glass-ceramics. Variation in the mean crystal size from 5 to 200 μm had no significant effect on the loss tangent. In these materials the volume fraction of the crystal phase, lithium disilicate was approximately constant since they all received the same crystallisation heat-treatment; the variations in mean crystal size were achieved by variation of the nucleation temperature.

From what has been stated earlier, it will be clear that the volume fraction of crystalline phase will be an important factor with regard to dielectric loss since it will determine the amount and composition of the residual glass phase. Experimental confirmation of this was given by Atkinson and McMillan (1976) who found for a lithia containing glass-ceramic that minimum values of loss tangent were obtained by heat-treating the material at 750°C, a temperature which also gave rise to the maximum volume fraction of crystalline phase.

Although the usual mechanisms of dielectric loss in glass-ceramics are related to ionic movements, in special cases other processes may be important. A study was made by Monneraye *et al.* (1968) of glass-ceramics prepared from a glass containing about 20 per cent TiO_2. It was found that the uncrystallised glass had a loss tangent of less than 20×10^{-4} at a frequency of 1 MHz. However, the glass-ceramic prepared from this glass exhibited an extremely high loss tangent of the order of 3000×10^{-4}. This very high value was attributed to semiconduction owing to non-stoichiometry in the form of an oxygen deficiency in titanate crystals present in the glass-ceramic. It was known for titanate crystals that incorporation of trivalent ions leads to a decrease of the conductivity and accordingly the effect of including a small proportion of chromic oxide (contributing Cr^{3+} ions) was investigated. The inclusion of as little as 0·04 mol. per cent Cr_2O_3 had the effect of reducing the loss tangent from 3300×10^{-4} to 18×10^{-4}. In other work it was shown that small amounts of Fe_2O_3 had similar effects in leading to reduced loss tangents for glass-ceramics containing titanate crystals.

Investigations of this nature emphasise the need for careful control of the state of oxidation of the parent glass in cases where there is a possibility that non-stoichiometric crystals can be formed.

The dielectric losses can increase noticeably as the temperature is raised, especially at the lower frequencies where alkali metal ion migration effects are important. At high frequencies ($\sim$ 1000 MHz) increase of temperature causes a less significant effect. These features are illustrated in Figs. 69, 70 and 71 for different types of glass-ceramic. In the last of these the curves are for a nominally alkali-free material but presumably a sufficient concentration of alkali metals occurs, possibly present as impurities, to account for the observed behaviour.

Some alkali-free glass-ceramics show a very small temperature-dependence of loss tangent at a frequency of 10 000 MHz. For example, glass-ceramics of

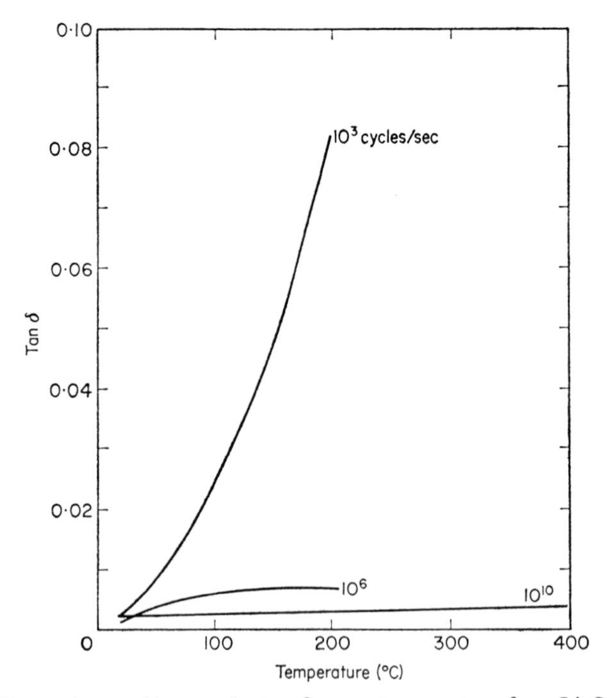

FIG. 69. Dependence of loss angle, tan δ, upon temperature for a Li_2O–ZnO–SiO_2 type glass-ceramic at various frequencies.

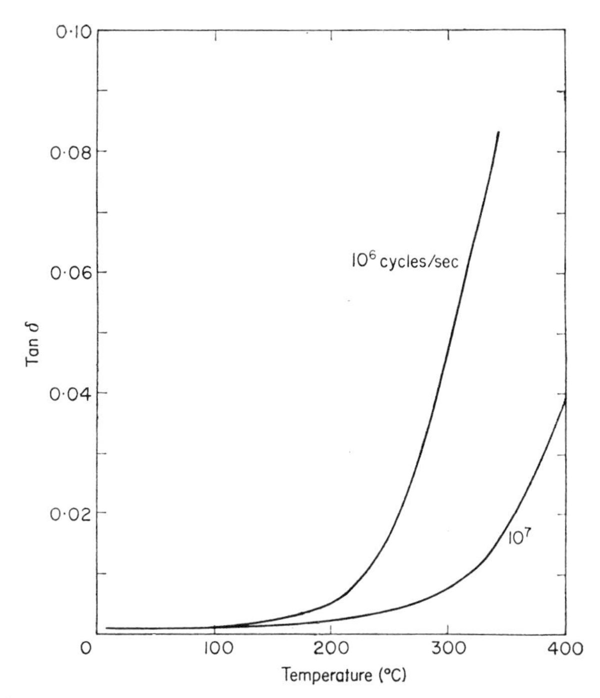

FIG. 70. Dependence of loss angle upon temperature for a Li_2O–ZnO–PbO–SiO_2 glass-ceramic.

the ZnO–Al_2O_3–SiO_2 type containing additions of BaO or SrO, had loss tangents at 400°C of only $\sim 8 \times 10^{-4}$.

b. Permittivity. The permittivity or dielectric constant represents the ratio of the electrical energy of the field set up in a dielectric material to that set up in a vacuum. As we have seen, the permittivities of many glass-ceramics at room temperature lie between 5 and 6 (Table XXVIII) and the values are not greatly affected by the frequency at which they are measured. At lower frequencies, the permittivity increases as the temperature is raised although the change in the value of the permittivity does not become appreciable until the temperature exceeds 150°C or so. At very high frequencies (10^9 to 10^{10} Hz) the value of the permittivity is scarcely affected by increase of temperature up to 400 to 500°C. These features are clearly shown by Figs 72 and 73.

A low permittivity combined with a low power factor is necessary if a material is to be used for insulation at high frequencies. In some cases it is required that the permittivity should change as little as possible with increase of temperature.

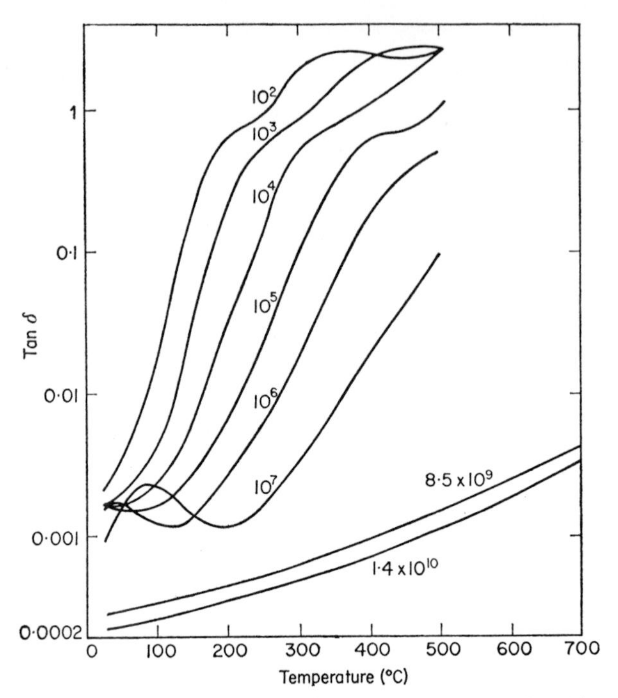

FIG. 71. Dependence of loss angle, tan δ, upon temperature for glass-ceramic no. 9606 at various frequencies.

The increases of permittivity at the lower frequencies are related to the increasing contributions from ion mobility and crystal imperfection mobility. Also, at higher temperatures dc conductivity effects become important. The combined effect is to give a sharp increase of permittivity.

There are two contributions to the change or permittivity with temperature and in a simplified form the temperature coefficient $\triangle K$ can be written:

$$\triangle K = Af(K) + D \tan \delta$$

where K is the permittivity, D is a constant and A is a quantity that expresses the influence of temperature on the polarisability of a particular material. For ionic crystals having $K < 20$, A is positive and for materials having $K > 20$, A is negative.

To achieve a glass-ceramic with a low temperature coefficient of permittivity, a low value of tan δ is clearly necessary. Also if a suitable proportion of crystals is present for which A is negative, there is the possibility of achieving a material with a zero temperature coefficient when $Af(K) = D \tan \delta$.

The permittivity-temperature curves in Fig. 74, determined at a frequency

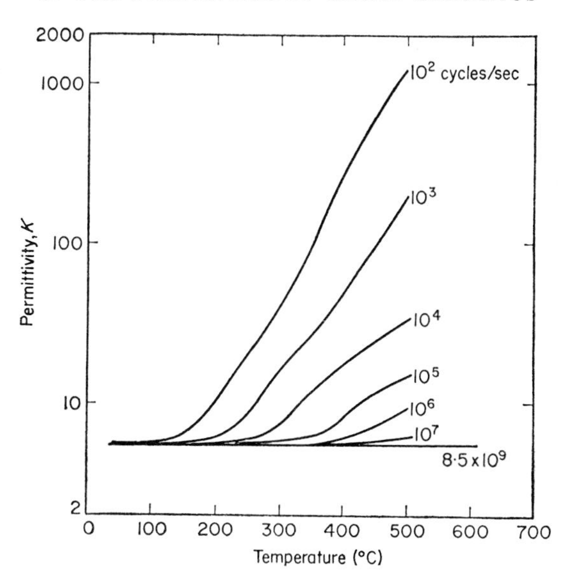

FIG. 72. Dependence of permittivity, K, upon temperature for glass-ceramic no. 9606 at various frequencies.

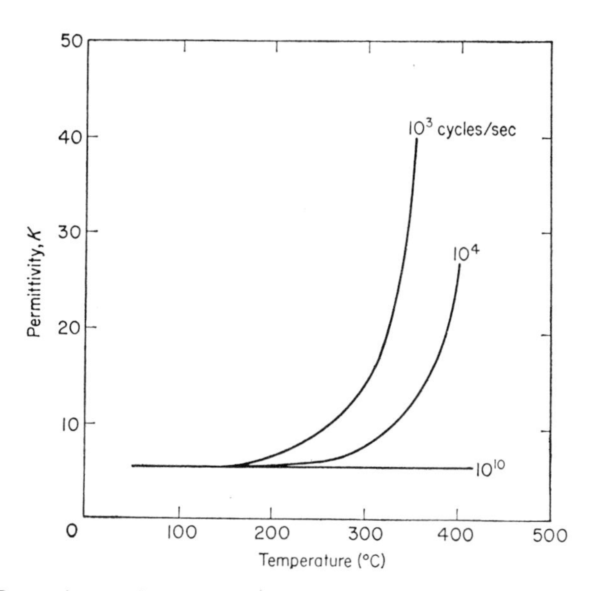

FIG. 73. Dependence of permittivity, K, upon temperature for a Li_2O–ZnO–SiO_2 glass-ceramic at various frequencies.

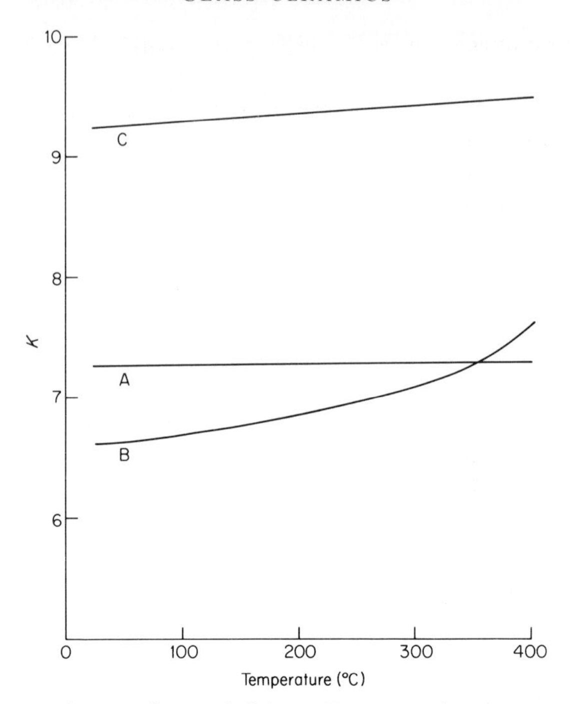

FIG. 74. Dependence of permittivity, K, upon the temperature for: A, $ZnO–Al_2O_3–SiO_2$ glass-ceramic; B, $ZnO–Na_2O–Al_2O_3–SiO_2$ glass-ceramic; C, 95 per cent alumina ceramic (measurement frequency 10^4 MHz).

of 10 000 MHz, illustrate the effect of reducing the value of tan δ. Curves A and B are for glass-ceramics having essentially the same crystallographic structure but the material represented by curve A is alkali-free whereas that represented by curve B contains 2 per cent Na_2O. The alkali-free material shows only a very small increase of permittivity up to 400°C and is in fact superior to a 95 per cent alumina ceramic (curve C) in this respect. The inferior performance of the Na_2O-containing glass-ceramic can largely be attributed to its much higher value of tan δ (150×10^{-4}) at 10 000 MHz as compared with that of the alkali-free material (18×10^{-4}). For glass-ceramics of the $MgO–Al_2O_3–SiO_2$ type containing cordierite as the major crystal phase, small temperature coefficients of permittivity can be achieved; between 20 and 400°C at a frequency of 10 000 MHz the change of permittivity is less than 4%. $ZnO–Al_2O_3–SiO_2$ glass-ceramics containing BaO also show small temperature coefficients since the increase of permittivity between 20 and 400°C is only about 3 per cent.

The relatively low values of permittivity for many glass-ceramics do not render them suitable for the production of capacitators. For this reason special glass-ceramics possessing high permittivities have been developed (Stookey, 1962).

Glass-ceramics having high permittivities, low power factors, high dielectric breakdown strengths and good insulation resistance can be produced from glassses which precipitate ferroelectric phases during suitable heat-treatment cycles. The phases in question are mainly of the perovskite type having the general formula ABO_3. The A ions are selected from the first, second and third groups of the periodic table and the B ions from the second to the fifth groups. In addition to barium titanate, other important compounds of this type are sodium niobate, cadmium niobate and other niobates; tantalates and zironates can also be present as ferroelectric compounds in some of the glass-ceramics.

The oxide constituents of these ferroelectric compounds do not form glasses and it is therefore necessary to include other oxides which will allow glass formation followed by controlled crystallisation. The glass-forming constituent, which is usually silica together in some cases with aluminium oxide, is introduced in the minimum amount possible in order to avoid excessive dilution of the ferroelectric phase.

Table XXIV gives some examples of the properties of high permittivity glass-ceramics and shows that materials having permittivities covering a wide range of values can be made.

Layton and Herczog (1969) have described some of the factors in the production of these materials. For example, variation of the silica content of $Na_2O–Nb_2O_3–SiO_2$ glasses from about 14 to 27 per cent causes a progressive increase of the grain size of the glass-ceramics from 0·02 to 1 μm. As shown by

TABLE XXIV

PROPERTIES OF GLASS-CERAMICS CONTAINING FERROELECTRIC CRYSTAL PHASES MEASURED AT 25°C AND AT 1 KHz

High permittivity crystal phases	Permittivity	Power factor, tan δ (per cent)
Barium titanate	30–2000	0·5–4·0
Sodium niobate cadmium niobate	375–590	2·1–2·4
Tungstic oxide	2100	120
Sodium niobate lead niobate	214	1·4
Barium zirconate, lead zirconate, cadmium zirconate	161	1·0

Herczog (1964), the permittivity of barium titanate glass-ceramic increases with increase of grain size (Fig. 75). Also at a grain size of 0·2 μm the peak in the dielectric constant-temperature curve disappears; this peak represents the ferroelectric-para-electric transition. Similar behaviour has been reported for colloidal barium titante but there is no generally accepted explanation. Another factor which affects the permittivity is of course the volume fraction of the ferroelectric phase. At 60 to 80 per cent volume fraction of this phase the permittivity of the glass-ceramic is about one third to one half of the permittivity of the pure crystal.

High permittivity glass-ceramics in which the crystals are an order of magnitude less in size than the wavelengths of visible light are transparent. The crystals in such materials generally have sizes in the range 5 to 50 nm. Layton and Herczog (1969) described materials in which the ferroelectric phases included barium titanate, sodium niobate and mixed niobates containing various monovalent and divalent cations. Kokubo and Tashiro (1976) described transparent glass-ceramics containing lead titanate as the ferroelectric phase.

These transparent high permittivity glass-ceramics show an electro-optical

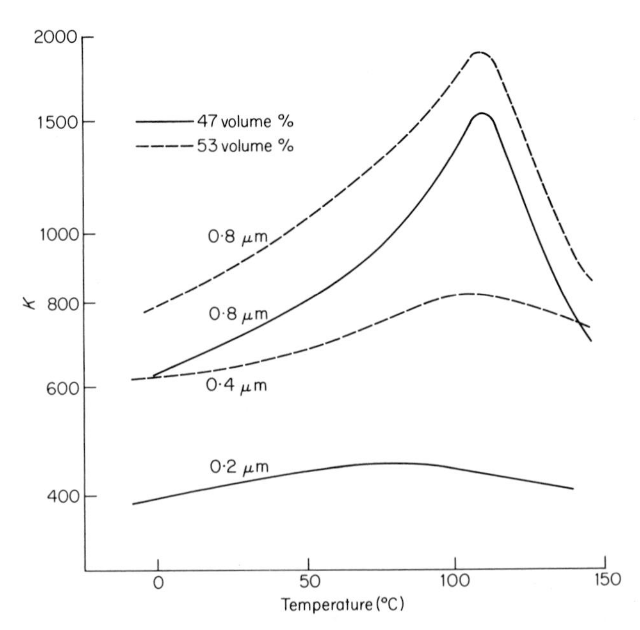

FIG. 75. Effects of grain size and temperature of the permittivity of a ferroelectric glass-ceramic (after A. Herczog, 1964, reproduced by courtesy of the American Ceramic Society).

effect in that when an electric field is applied, they rotate the plane of polarisation of a beam of plane-polarised light transmitted through them. Generally a quadratic electro-optical effect is produced in which the induced birefringence $\triangle n$ is proportional to the square of the applied field E:

$$\triangle n = \tfrac{3}{4} n^3 g K^2 E^2$$

where n is the refractive index, g is the electro-optic coefficient and K is the dielectric constant of the glass-ceramic.

c. Dielectric strength. If an insulating material is subjected to a large voltage gradient, failure of the material may occur due to puncture. This possibility is one of the important factors which govern the design of high-voltage insulators. Dielectric break-down strength is also of importance in the design of capacitors required to operate at high voltages such as those used for energy storage applications.

Dielectric breakdown of insulating materials can take place in two ways. The first, sometimes referred to as an intrinsic dielectric breakdown, is of electronic origin and the second process, sometimes described as thermal breakdown, results from local overheating arising from electrical conduction; this local conduction increases to a point where actual puncture of the dielectric can occur. In practice it is very difficult to separate the two effects and even with elaborate experimental techniques the values obtained fall short of the actual intrinsic strength of the material. In addition to these difficulties, it is found that the experimental values obtained are strongly dependent upon the conditions of the test. Factors such as the size and shape of electrodes, the thickness of the test material subjected to the electrical stress, the rate of application of the voltage, and the duration of the applied voltage may all influence the values measured. If the breakdown strength is being determined under alternating-current conditions, the frequency has a marked influence since lower breakdown strengths are observed at higher frequencies. Clearly, therefore, the conditions of test must be closely specified and specimens of standard design must be used. The tests are often carried out under oil of specified quality to suppress corona and discharges and the breakdown strength at power frequencies (50 or 60 Hz) is often of the greatest interest.

Glass-ceramics have high dielectric breakdown strengths compared with those measured for conventional ceramics. This is illustrated by the data given in Table XXV. These results were obtained for specimens of the dimension given in the German Standard DIN 40685 at a measurement frequency of 50 Hz. This design of specimen has been found to give more reproducible results than other types, such as that proposed in A.S.T.M. D116–44. The DIN specimen is designed to prevent flashover and the test thickness of the

TABLE XXV

DIELECTRIC BREAKDOWN STRENGTHS OF GLASS-CERAMICS AND CONVENTIONAL CERAMICS USING SPECIMENS TO DIN 40685: TEST FREQUENCY 50 Hz; TEMPERATURE 20°C

Material	Dielectric breakdown strength (kV/mm)
Glass-ceramics	
Lithium-zinc-silicate (ZnO content c.5%)	47
Lithium-zinc-silicate (ZnO content c.30%)	38
Lithium-zinc-lead-silicate	28
Conventional ceramics	
High voltage electrical porcelain	25
High alumina ceramics (95% Al_2O_3)	14–23

dielectric is 1·5 mm. The high values of dielectric strength for the glass-ceramics as compared with those for conventional ceramics may be partly due to the complete absence of closed pores in the former materials. Porosity tends to give variations in the local electrical field thus giving rise to low measured values. The very homogeneous and fine-grained nature of the glass-ceramics may be another factor which favours high breakdown strengths.

D. THERMAL PROPERTIES

1. *Thermal Expansion Characteristics*

a. General considerations. The dimensional changes which occur with change of temperature are of great importance from a number of points of view. For example, if a glass-ceramic is required to have high thermal shock resistance, the coefficient of thermal expansion should be as low as possible to minimise strains resulting from temperature gradients within the material. Also if the glass-ceramic is to be sealed or otherwise rigidly joined to another material such as a metal, close matching of the thermal expansion coefficients is necessary to prevent the generation of high stresses when the composite article is heated or cooled. In some applications, dimensional stability with change of temperature may be important and a glass-ceramic of near zero thermal expansion coefficient would be required for this.

Glass-ceramics are remarkable for the very wide range of thermal expansion coefficients which can be obtained. At one extreme, materials having negative coefficients of thermal expansion are available while for other compositions very high positive coefficients are observed. Between these two extremes there exist glass-ceramics having thermal expansion coefficients practically equal to zero and others whose expansion coefficients are similar to those of ordinary glasses or ceramics or to those of certain metals or alloys.

b. Influence of chemical constitution and crystal types upon thermal expansion characteristics. It was shown earlier that the thermal expansion coefficient of a glass-ceramic can be markedly different from that of the parent glass. The process of crystallisation introduces phases having coefficients of thermal expansion usually different from that of the parent glass and the glass-ceramic may have a higher or lower coefficient of expansion depending on the crystal phases formed. It should be borne in mind that a glass-ceramic is a composite material and its thermal expansion coefficient is a function of the thermal expansion coefficients and elastic properties of all the phases present, including the residual glass. Thus while the development of crystal phases usually causes the major changes in the expansion coefficient, crystallisation will alter the composition of the residual glass phase from that of the parent glass and this must be taken into account when attempting to analyse the thermal expansions of glass-ceramics in relation to their constitution. In certain cases, the changes in composition of the residual glass phase result in changes of the thermal expansion of a sufficient magnitude to counterbalance the effects of the development of the crystal phase. Consequently, the glass-ceramic and the parent glass have closely similar thermal expansion coefficients. Such cases are fairly rare; the more general behaviour is for the crystallisation process to result in a distinct change of thermal expansion coefficient.

We expect a composite material such as a glass-ceramic to follow the thermal behaviour discussed by Kingery (1960) who gives the following expression for thermal coefficient:

$$\alpha = \frac{\dfrac{\alpha_1 K_1 F_1}{\rho_1} + \dfrac{\alpha_2 K_2 F_2}{\rho_2} + \ldots\ldots\ldots\ldots}{\dfrac{K_1 F_1}{\rho_1} + \dfrac{K_2 F_2}{\rho_2} + \ldots\ldots\ldots\ldots}$$

where α_1 α_2 are the expansion coefficients, K_1 K_2 are the bulk moduli and F_1 F_2 are the weight fractions of the components. If the Poisson's ratios are similar the moduli of elasticity can be substituted for the bulk moduli. This expression assumes ideal behaviour with the contraction of each component equalling the overall contraction and all microstresses being pure hydrostatic tension or compression.

Knowing the volume fractions of the crystal phases and their physical properties it is therefore possible to estimate the thermal expansion coefficient of a glass-ceramic. One of the main difficulties is in making the appropriate allowance for the changes in properties of the residual glass phase. The composition of this phase can be estimated, however, if the volume fractions of crystal phases are known with reasonable accuracy from electron microscopy or quantitative diffraction studies. With this information, the approximate properties of the glass phase can be derived from additive relationships.

The wide range of thermal expansion of crystals that can be developed in glass-ceramics is illustrated by the data given in Table XXVI which is largely based on data given by Stutzman *et al.* (1959). The values are for polycrystalline aggregates so that anisotropic effects (discussed on p. 184) are averaged out. Since the crystals in glass-ceramics are normally randomly orientated, the values given may be taken as representative of the contributions of the various crystal phases to glass-ceramic thermal expansion coefficients. Since the temperature ranges are not identical in all cases, these are quoted together with the expansion figures.

An extremely wide range of thermal expansion coefficients is covered by the different crystal types and the development of these phases in appropriate proportions forms the basis of the production of glass-ceramics having controlled thermal expansion coefficients. Thus a low expansion glass-ceramic may contain beta-eucryptite, beta-spodumene or cordierite as major phases while at the other extreme, a high expansion glass-ceramic may contain major proportions of crystals such as lithium disilicate, quartz or cristobalite.

It should be pointed out that crystal phases at the extremes of the ranges are likely to exhibit a large mismatch of their expansion coefficients with that of the residual glass phase. Taking an extreme case, no glasses having negative thermal expansion coefficients in the temperature range above 20°C are known and therefore beta-eucryptite must always be mismatched in expansion coefficient to the residual glass phase. It is likely that mismatching of expansion also occurs for other low expansion phases. The effect of these unmatched thermal expansions will be to cause microstresses to arise in the glass-ceramic. In extreme cases these could result in microcracking although the situation is alleviated to some extent by the generally small sizes of the crystal phases so that the localised stresses are normally restricted to

TABLE XXVI

THERMAL EXPANSION COEFFICIENTS OF CRYSTAL TYPES WHICH MAY BE PRESENT IN GLASS-CERAMICS

Crystal phase	Thermal expansion coefficient (°C^{-1})
$Li_2O.Al_2O_3.SiO_2$ (Beta-eucryptite)	-86×10^{-7} (20–700°C) -64×10^{-7} (20–1000°C)
$Al_2O_3.TiO_2$ (Aluminium titanate)	-19×10^{-7} (25–1000°C)
$2MgO.2Al_2O_3.5SiO_2$ (Cordierite)	6×10^{-7} (100–200°C) 26×10^{-7} (25–700°C)
$Li_2O.Al_2O_3.4SiO_2$ (Beta-spodumene)	9×10^{-7} (20–1000°C)
$BaO.Al_2O_3.2SiO$ (Celsian)	27×10^{-7} (20–100°C)
$CaO.Al_2O_3.2SiO_2$ (Anorthite)	45×10^{-7} (100–200°C)
$MgO.SiO_2$ (Clinoenstatite)	78×10^{-7} (100–200°C)
$MgO.TiO_2$ (Magnesium titanate)	79×10^{-7} (25–1000°C)
$2MgO.SiO_2$ (Forsterite)	94×10^{-7} (100–200°C)
$CaO.SiO_2$ (Wollastonite)	94×10^{-7} (100–200°C)
$Li_2O.2SiO_2$ (Lithium disilicate)	110×10^{-7} (20–600°C)
SiO_2 (Quartz)	112×10^{-7} (20–100°C) 132×10^{-7} (20–300°C) 237×10^{-7} (20–600°C)
SiO_2 (Cristobalite)	125×10^{-7} (20–100°C) 500×10^{-7} (20–300°C) 271×10^{-7} 20–600°C)
SiO_2 (Tridymite)	175×10^{-7} (20–100°C) 250×10^{-7} 20–200°C) 144×10^{-7} (20–600°C)

acceptable levels. Nevertheless, in formulating a glass-ceramic, which is to contain low expansion crystal phases, it is prudent to avoid the inclusion of constituents which tend to concentrate in the residual glass phase and to raise its coefficient of thermal expansion. Thus excess concentrations of lithium oxide over that required to form low expansion lithium aluminosilicate crystals or the presence of significant concentrations of other alkali metal oxides which do not take part in the crystalline phases are generally to be

TABLE XXVII
THERMAL EXPANSION COEFFICIENTS AND CRYSTAL PHASES PRESENT IN GLASS-CERAMICS

SiO_2	Al_2O_3	ZnO	MgO	CaO	Li_2O	Na_2O	K_2O	TiO_2	P_2O_5	Ref.	Thermal expansion coefficient $\times 10^7$ ($°C^{-1}$)	Crystal phases
57·2	28·7	—	2·7	—	8·4	—	—	—	3·0	3	−38·7	β-Eucryptite
68·0	21·4	—	—	—	3·8	—	—	5·7	—	1	−9·8	β-Eucryptite
54·5	34·5	—	—	—	5·5	—	—	5·5	—	1	1·1	β-Eucryptite
70·7	18·1	—	—	—	2·6	—	—	4·8	—	1	5·1	β-Spodumene, cordierite, rutile
45·5	30·5	—	12·5	—	—	—	—	11·5	—	2	14·1	Cordierite, rutile
65·5	21·0	—	—	—	9·0	—	—	4·5	—	1	14·5	β-Spodumene, rutile
45·8	25·3	—	17·8	—	—	—	—	11·1	—	2	22·6	Cordierite, magnesium titanate
52·5	26·5	—	11·9	—	—	—	—	9·1	—	2	28·3	Cordierite, rutile

57·8	8·9	—	22·2	—	—	—	—	11·1	—	2	39·9	Cordierite, rutile cristobalite
56·0	20·0	—	15·0	—	—	—	—	9·0	—	2	56·0	Cordierite, magnesium titanate, rutile cristobalite
58·1	19·1	—	13·7	—	—	—	—	9·1	—	2	63·3	Cordierite, cristobalite, magnesium titanate
65·7	11·2	—	—	—	18·6	—	—	—	4·5	3	75·0	β-Spodumene/silica S.S., lithium metasilicate
78·5	3·9	—	—	—	12·1	—	—	—	3·0	3	102	
82·3	—	—	3·7	—	11·0	—	—	—	3·0	3	145	Quartz, lithium disilicate
59·2	—	27·1	—	—	9·0	—	2·0	—	2·7	4	160	Cristobalite, lithium zinc silicate

References: 1. Stookey (1959), 2. Stookey (1960), 3. McMillan and Partridge (1963a), 4. McMillan and Partridge (1963b).
S.S. stands for solid solution.

avoided. A similar problem can arise when crystals of very high thermal expansion coefficient are developed; in this case the residual glass phase will almost certainly have a lower coefficient of expansion than the crystal phases. Following the principles outlined above, the aim in this case will be to ensure that the composition of the residual glass phase is such as to minimise the extent of thermal expansion mismatching.

The contribution of a particular crystal phase to the thermal expansion of a glass-ceramic may be modified if the crystal enters into solid solution with another crystal phase. Beta-eucryptite and beta-spodumene can form a series of solid solutions and these would have thermal expansion coefficients intermediate between those of the end members of the series. Beta-spodumene, in which the proportions of Li_2O, Al_2O_3 and SiO_2 are 1:1:4, can also form solid solutions with silica. F. A. Hummel (1951) has shown that while the thermal expansion coefficient of pure beta-spodumene is 9×10^{-7} (25–100°C) and that for the 1:1:6 solid solution is 5×10^{-7} that for the 1:1:8 solid solution is 3×10^{-7} and that for the 1:1:10 solid solution is 5×10^{-7}.

It will be noted that the thermal expansion coefficients of the crystalline forms of silica are strongly dependent upon the temperature range for which they are measured. This occurs because these crystals undergo changes of structure, accompanied by changes of thermal expansion coefficient, at characteristic temperatures. The effects of these phase inversions upon the thermal expansion characteristics of glass-ceramics are discussed more fully on pp. 229–30.

Table XXVII gives thermal expansion coefficients for a selection of glass-ceramics and illustrates in a general way the influence of various crystal phases upon thermal expansion coefficient. The volume fraction of a particular crystal phase present in the glass-ceramic is of course important as well as its thermal expansion properties so that glass-ceramics containing the same crystal phases can have different thermal expansion coefficients. In designing glass-ceramics to have specified thermal expansion coefficients, therefore, it is possible to vary not only the types of crystals but also their proportions in order to achieve the desired properties.

The effects of systematic changes of the types of crystal phases in a glass-ceramic are illustrated in Fig. 76. The replacement of Al_2O_3 by ZnO in this material at first has only a relatively small effect upon the thermal coefficient but in the region where the molecular ratio of ZnO equals that of Al_2O_3 there is a rapid increase of the coefficient followed by a more gradual rise for higher ZnO contents. These effects are caused by changes in the crystal phases present in the glass-ceramics. For the materials containing less ZnO than Al_2O_3, the major phase is a beta-spodumene/silica solid solution but this disappears and is replaced by the high expansion phase, quartz when the ZnO content exceeds the Al_2O_3 content. A progressive increase in the amount of

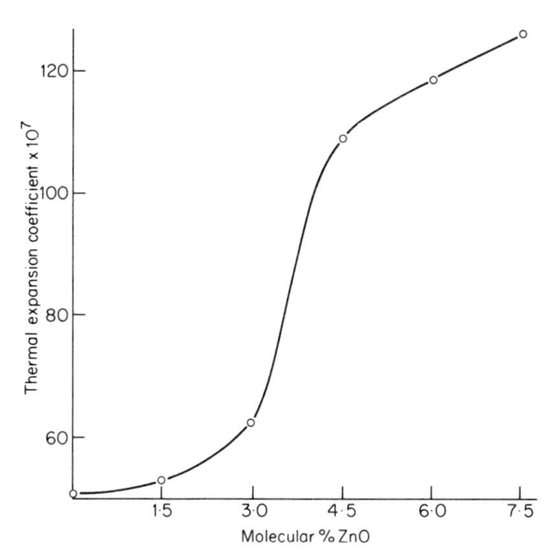

FIG. 76. Effect on the thermal expansion coefficient of a Li_2O–ZnO–Al_2O_3–SiO_2 glass-ceramic of substituting ZnO for Al_2O_3.

quartz accounts for the steady increase of expansion coefficient with the higher ZnO contents.

Figure 75 illustrates an important point. This is that comparatively small variations of composition in the region where the ratio $ZnO:Al_2O_3 \approx 1$ will result in large variations of thermal expansion coefficient. Such compositions would therefore be unsuitable from a practical viewpoint for the production of glass-ceramics having controlled thermal expansion coefficients. The composition ranges on either side of this region would, however, allow controlled thermal expansion coefficients to be realised. As a general rule, glasses in which small deviations of composition lead to large changes of the crystal content, and therefore of physical properties of the glass-ceramics would be avoided.

c. Effect of heat-treatment schedule upon thermal expansion characteristics. The thermal expansion characteristics of a glass-ceramic can be markedly affected by the heat-treatment schedule since this determines the proportions and nature of the crystal phases present. This fact is strikingly illustrated by comparison of curves B and D in Fig. 77. Both of these glass-ceramics were prepared from the same parent glass composition so that chemically they are identical. In one case, however, the heat-treatment schedule was adjusted to produce all of the crystalline silica in the form of cristobalite (curve B) and in the other case to give the silica entirely in the form of quartz. By variation of the heat-treatment process, glass-ceramics containing mixtures of quartz and

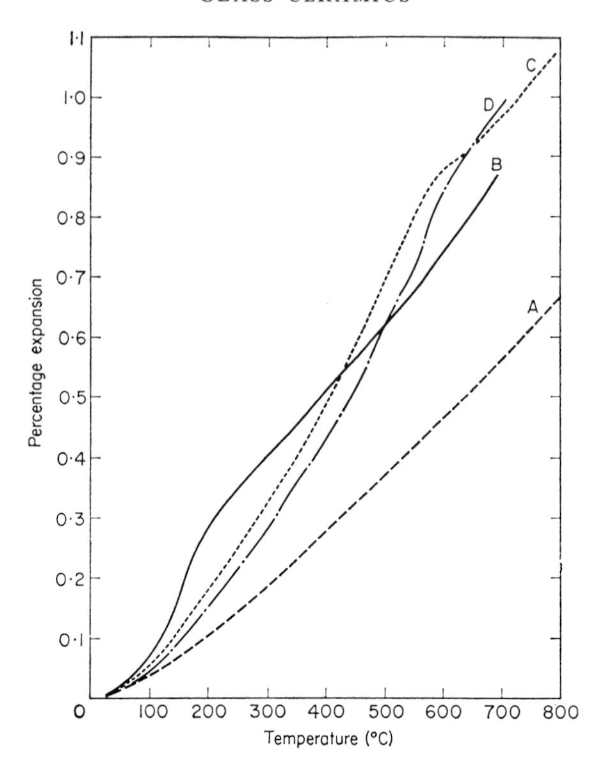

FIG. 77. Thermal expansion curves of glass-ceramics showing influence of phase inversions.

cristobalite in various proportions could also be produced from the same glass.

Variations of the heat-treatment schedule can also result in changes in the volume fractions of crystal phases with consequent effects upon the thermal expansion coefficients of glass-ceramics. This effect was studied by McMillan *et al.* (1966) who measured the thermal expansion coefficients of glass-ceramics produced from a phosphate-nucleated Li_2O–ZnO–SiO_2 glass-ceramic heat-treated under various conditions. Figure 78 shows the effect of varying the upper heat-treatment temperature for specimens that had all received a nucleation treatment of 1 hour at 500°C. Initially, lithium disilicate crystals are precipitated and the specimen heat-treated at 700°C contains approximately equal volume fractions of this phase and residual glass. Heat-treatment at temperatures up to 700°C causes a slight fall in the thermal expansion even though the lithium disilicate crystals have a higher thermal expansion coefficient than the parent glass ($\sim 83 \times 10^{-7}\,°C^{-1}$). This is attributed to depletion of the glass phase of lithium oxide which would cause a

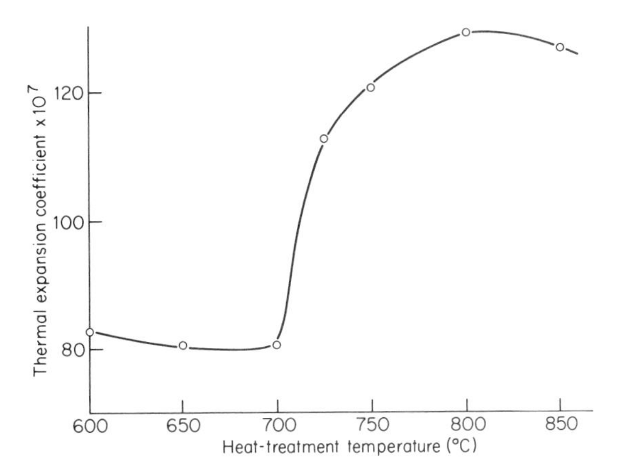

FIG. 78. Effect of varying the crystallisation heat-treatment temperature upon the thermal expansion coefficient of a Li_2O–ZnO–SiO_2 glass-ceramic.

reduction of its thermal expansion coefficient and this results in a small net reduction in the expansion coefficient of the glass-ceramic. Heat-treatment above 700°C causes a marked increase of the thermal expansion coefficient and this is attributed to the crystallisation of the high silica glass phase resulting in the production of quartz crystals which have a high coefficient of expansion. The material heat-treated at 800–850°C for 1 hour contains approximately equal amounts of lithium disilicate and quartz with a small volume fraction (about 5 per cent) of residual glass phase.

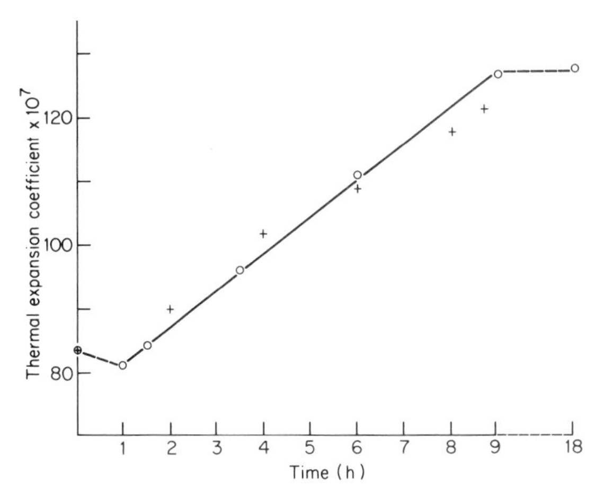

FIG. 79. Effect of isothermal treatment at 700°C upon the thermal expansion coefficient of a Li_2O–ZnO–SiO_2 glass-ceramic, ○, experimental values; +, calculated values.

Isothermal treatment at 700°C of specimens nucleated for 1 hour at 500°C gives the results shown in Fig. 79. The steady increase of expansion coefficient for heat-treatment durations greater than 1 hour is attributed to the progressive development of the quartz phase; this process appears to be complete after heat-treatment for 9 hours. Making the assumptions that after heat-treatment for 1 hour the material contained 50 per cent lithium disilicate and 50 per cent glass and after 9 hours 50 per cent lithium disilicate and 50 per cent quartz it was possible to calculate thermal expansion coefficients for various heat-treatment times. As shown in Fig. 79 the agreement between the calculated points and the experimentally derived curve is fairly good and gives support to the idea that thermal expansion coefficients of glass-ceramics might be predictable from a knowledge of their crystallographic constitution.

2. *Refractoriness*

a. General considerations. The temperature to which a glass-ceramic can be heated without exhibiting deformation can be important if the material is required to operate or to be processed at high temperatures. For this reason, it is necessary to know how glass-ceramics compare with other ceramic-type materials with regard to refractoriness and it is also useful to relate this characteristic to the glass-ceramic type.

If a glass-ceramic specimen is heated, a temperature will be reached at which it deforms and if the heating is continued the specimen will, of course, melt. The temperature at which deformation is observed depends upon the rate of heating, upon the the method of supporting the specimen and upon the load applied to the specimen. Standardization of the conditions of test is therefore necessary. In one method a bar specimen is supported on knife edges placed at a specified distance apart and heated at a constant rate; the deformation temperature is established by noting when sagging of the bar specimen occurs. In another method, which is very convenient, the maximum point attributed on the thermal expansion curve is recorded since above a certain temperature the specimen no longer expands due to the occurrence of softening; the temperature observed in this manner is described as the dilatometric softening temperature.

Measurement of the creep rate of standard specimens under controlled conditions is of great value since this information can be used to make deductions concerning the basic kinds of process responsible for deformation.

The factors which govern the deformation of a glass-ceramic are complex, since there are one or more crystal phases present together with a residual glass phase; this phase is probably distributed fairly uniformly between the crystals. It is very likely that it is the characteristics of the glass-phase which largely govern the refractoriness of a glass-ceramic since most of the crystal phases present have high melting temperatures. Obviously, the proportion of glass-phase present will have a strong bearing on the refractoriness of the glass-ceramic, and from this point of view it is desirable to limit the glass-phase to the smallest possible proportion. In addition, the composition of the glass-phase is of great importance and it is necessary to ensure that oxides which will lower the softening temperature of the glass-phase are present only in small proportions. Oxides which would have this effect include the alkali metal oxides, boric oxide and lead oxide. In addition to the possibility of viscous flow of the glass-phase as a cause of deformation, there is also the possibility that the crystal phases may begin to redissolve as the temperature increases. Heating to a sufficiently high temperature will, of course, reconvert the glass-ceramic to its initial liquid or glass-like state.

b. Relationship between glass-ceramic type and refractoriness. Glass-ceramics of the Al_2O_3–SiO_2 type described by MacDowell and Beall (1969) are representative of highly refractory materials that can be produced by crystallisation of glasses. These contain mullite, $3Al_2O_3.2SiO_2$, as the major phase. The high melting point of this compound (1850°C) is noteworthy but the highly refratory nature of the residual glass phase in these materials is also likely to contribute to their high temperature dimensional stability. Some of these materials were heat-treated at 1500°C for periods of 10 hours without apparently exhibiting significant distortion.

The incorporation of MgO to give glass-ceramics in which cordierite appears as a major phase, while resulting in materials less refractory than the binary Al_2O_3–SiO_2 glass-ceramics, nevertheless enables materials having valuable high temperature properties to be produced. These have the advantage of being easier to prepare and shape as compared with the Al_2O_3–SiO_2 materials. Depending on the Al_2O_3 content, the deformation temperatures of these materials range from 1275°C to 1370°C as reported by Stookey (1960). Minor constituents have a marked effect upon the refractoriness of these materials.

The introduction of various oxides into a glass-ceramic having the weight percentage composition SiO_2; 43·5; Al_2O_3: 17·4; TiO_2: 8·7; ZnO: 26·1; RO: 4·3 where RO was MgO, CaO, BaO or PbO resulted in glass-ceramics having deformation temperatures as given in Table XXVIII. It should be noted that the reduction of deformation temperature is greater, the larger is the radius of the metal cation introduced. If the substitution had been made on a molecular rather than a weight basis, the effects of increasing ionic radius would have been more noticeable. The effects of the various oxides in lowering glass viscosity would show the same pattern, with lead oxide having the greatest effect. This suggests that the effects of the oxides on the refractoriness of the glass-ceramics arises from their incorporation in the residual glass phase.

TABLE XXVIII

EFFECT OF MINOR (4·3 WEIGHT PER CENT) CONSTITUENTS ON DEFORMATION TEMPERATURES OF GLASS-CERAMICS

Oxide	Deformation temperature (°C)
MgO	1125
CaO	1100
BaO	1025
PbO	875

Although in this case lead oxide resulted in the lowest deformation temperature, some glass-ceramics containing appreciable proportions of this oxide can have quite high deformation temperatures. For example, $PbO–Al_2O_3–SiO_2–TiO_2$ compositions having deformation temperatures ranging from 750°C to 1150°C have been reported (Stookey, 1960). These compositions contained 32 to 46 per cent PbO and the refractoriness diminished as the Al_2O_3 content decreased. In compositions of this type a high proportion of the lead oxide may be incorporated in crystalline phases such as lead titanate.

The data given in Table XXIX show effects of composition upon the dilatometric softening points of a number of glass-ceramics. Compositions 1 to 5 contained a small amount of gold and compositions 6 to 11 a small proportion of a metallic phosphate to act as nucleation catalysts. It is clear from these data that incorporation of aluminium oxide into a lithia-silica composition increases the refractoriness and is more effective in this respect than magnesium oxide. Other effects are that increasing the lithia content lowers the refractoriness of the glass-ceramics and that the magnesium-aluminosilicate composition is considerably more refractory than the molecularly equivalent lithium-aluminosilicate composition.

Glass-ceramics are more refractory than most conventional glasses since, although some aluminosilicate glasses may have dilatometric softening temperatures of 800°C or so, and fused silica glass is even more refractory, commonly available glasses have dilatometric softening points not exceeding 500 to 600°C. Thus glass-ceramics are more suitable for high-temperature applications than most glasses. Comparison with conventional ceramics is not too easy since the methods of assessing refractoriness often differ from those which have been used for glass-ceramics. Nevertheless a qualitative comparison can be made and it appears that while glass-ceramics are generally less refractory than high alumina ceramics, the more refractory glass-ceramics, chiefly those free from alkali metal oxides, have deformation temperatures approaching those of steatite or forsterite ceramics.

TABLE XXIX

DILATOMETRIC SOFTENING TEMPERATURE OF GLASS-CERAMICS AND PARENT GLASSES

Composition no.	Molecular percentage composition					Softening temperatures (°C)	
	SiO_2	Al_2O_3	ZnO	MgO	Li_2O	Parent glass	Glass-ceramic
1	71	—	—	—	29	500	850
2	71	6	—	—	23	530	980
3	71	6	—	6	17	550	950
4	71	6	6	—	17	575	970
5	65	6	6	—	23	560	960
6	75	—	—	5	20	560	910
7	70	—	—	10	20	530	895
8	75	5	—	—	20	515	>1000
9	70	10	—	—	20	620	>1000
10	50	10	—	—	40	510	920
11	50	10	—	40	—	805	>1000

c. Viscous deformation and creep resistance. Deformation of glass-ceramics under stress at high temperatures is technologically important in connection with engineering applications such as heat exchangers and other components of gas turbine engines. A knowledge of the factors underlying creep behaviour is therefore essential. For glass-ceramics, we may expect that the residual glass will exert a strong influence because this phase is capable of undergoing viscous deformation.

For a large number of materials, the curves of deformation versus time can be divided into three fairly well defined regimes:

(*i*) Primary creep which starts at the instant of loading and diminishes with time.

(*ii*) Secondary creep in which a steady creep rate is sustained for a significant period.

(*iii*) Tertiary creep in which the creep rate rises drastically leading to specimen failure.

Secondary creep is considered to correspond with a region of constant internal "creep structure". If this form of creep is not found for a given material it is believed that the creep structure changes continuously as a function of the overall strain and never attains equilibrium. The secondary creep rate is assumed to be a separable function of three independent variables, temperature T, stress σ and structure factor s. The structure factor is assumed to include all structural variables which affect creep such as grain size, dislocation density, porosity. The creep rate, $\dot{\varepsilon}$, can thus be represented as

$$\dot{\varepsilon} \propto f(T) f(\sigma) f(s)$$

At a fixed temperature and assuming the structure factor has attained equilibrium

$$\dot{\varepsilon} \propto f(\sigma)$$

The dependence of creep rate on stress is not linear and generally a relationship of the form

$$\dot{\varepsilon} = A\sigma^n$$

where A is a constant is found to hold. The value of the stress exponent, n, is characteristic of the deformation mechanism. Values of n close to unity are associated with a deformation process involving viscous flow, whereas higher values of about 3 to 4 are generally accepted as evidence of a nonviscous process (eg dislocation climb) as the limiting stage.

Since creep is thermally activated the temperature function is of the form

$$f(T) \propto \exp(-\triangle H/RT)$$

where $\triangle H$ is the activation energy for creep. For polycrystalline ceramic materials, values of 160–190 Kcal/mole have generally been obtained and it is generally accepted that these high activation energies are associated with non-viscous creep (dislocation movement) mechanisms, low activation energies for creep (90 to 130 Kcal/mole) are associated with viscous flow mechanisms.

A number of aspects of the tensile and compressive creep behaviour of a glass-ceramic of the Li_2O–ZnO–SiO_2 type have been investigated by Morell and Ashbee (1973). The crystalline phases present were lithium disilicate, lithium zinc silicate and quartz. The residual glass phase constituted about 20 vol per cent of the glass-ceramic and the weight percentage composition of this phase was estimated to be 20 K_2O, 80 SiO_2. If this glass phase were the rate controlling phase, allowing relative sliding, it would be expected that the glass-ceramic would behave as a Newtonian fluid, demonstrating infinite extensibility, a stress exponent of unity and an activation energy characteristic of the glass phase.

The three characteristic regions of creep were observed. Primary creep was considered to be due to the delayed build-up of elastic strain in the crystalline phases controlled by the viscoelastic behaviour of the glass phase. Tertiary creep was caused in tension by "tearing" or crack generation from the surface, and in compression by void formation and shear failure. In the secondary creep regime, the activation energy was the same both in tension and in compression and was also the same as that for primary creep (about 130 to 160

Kcal/mole). It was concluded that the creep properties of this glass-ceramic are strongly influenced by the structural morphology and are rate controlled by the glass phase. Plastic deformation of the crystal does not contribute a significant effect but purely viscous flow with the stress exponent, n, equal to unity only occurred at low stresses. At higher stresses the value of n ranged upwards to 6. It was considered that this was due to the existence of high negative hydrostatic pressures at high stresses and these caused nucleation and growth of voids so that deformation was no longer at constant volume.

The void production may depend upon the nature of the applied stress field. In torsional creep the stress exponent is in the range 1·4 to 1·7 and does not rise significantly at high applied stress as shown by Heuse and Partridge (1974). These investigators studied several glass-ceramics and carried out comparison tests on a 95 per cent alumina ceramic. The least refractory material studied was a low alumina content Li_2O–Al_2O_3–SiO_2 composition containing lithium disilicate and quartz as major crystal phases. This material showed significant deformation under the test conditions at 750°C and the activation energy of creep was in the range 138 to 158 $Kcal/mole^{-1}$. A composition from the same system but having a higher alumina content and containing a β-spodumene solid solution as the major crystal phase was more refractory; for equivalent deformation the temperatures were about 100°C higher. The activation energy, however, was not greatly different being in the range 120 to 154 $Kcal/mole^{-1}$. Alkali-free glass-ceramics of the BaO–ZnO–Al_2O_3–SiO_2 and CaO–Al_2O_3–SiO_2 type showed even greater refractoriness. The latter only began to show significant deformation under constant load at temperatures above 1100°C. In fact its behaviour was closely similar to that of the 95 per cent alumina ceramic.

Studies of tensile creep of glass-ceramics of the molecular composition $Li_2O.Al_2O_3.4SiO_2$ nucleated by the inclusion of 3 molecular per cent TiO_2 were made by Bold and Groves (1978). By restricting the upper heat-treatment temperature to 1000°C a glass-ceramic containing a beta quartz solid solution as the principal crystal phase was produced. The use of an upper heat-treatment temperature of 1200°C allowed the development of beta spodumene as the crystal phase. Both materials were highly crystalline with little evidence of a residual glass phase; for the beta quartz material the mean grain size was 0·14 μm and for the beta spodumene material 1·0 μm. It was found for both materials that the activation energy governing strain rate was 695 KJ mol^{-1} and the stress exponent, n, was 1·4 for both. Microstructural studies of the materials after large creep strains suggested that the mechanism of creep involved grain boundary sliding accommodated by grain boundary diffusion. It was noted that despite the occurrence of large creep strains before failure, the ultimate fracture was brittle in character.

3. *Specific Heat*

The specific heats of glass-ceramics do not vary greatly from one composition to another and the mean specific heats (20 to 400°C) are generally in the range 0·22 to 0·28 cal g^{-1} $°C^{-1}$. Thus the values are not too dissimilar to those observed for glasses and ceramics. Insufficient data have been accumulated to permit the relationship between specific heat and the composition of a glass-ceramic to be explored, but for glass-ceramics containing lithia and silica as major constituents, experimental results suggest that the specific heat tends to decrease as the silica content is reduced. The specific heats of glass-ceramics increase as the temperature is raised and the relationship between the two parameters is similar to that observed for glasses and ceramics. Figure 80 shows the relationship between specific heat and temperature for two glass-ceramics.

4. *Thermal Conductivity and Thermal Diffusivity*

a. Thermal conductivity. A glass-ceramic may be required for use as a thermal conductor or as a thermal insulator and its suitability for these applications will be largely determined by the rate of heat transfer through it under a given temperature gradient. A reasonably high thermal conductivity is desirable if the material is to be used for cooking vessels, for example. In addition, the thermal shock resistance of the material is influenced by the thermal conductivity.

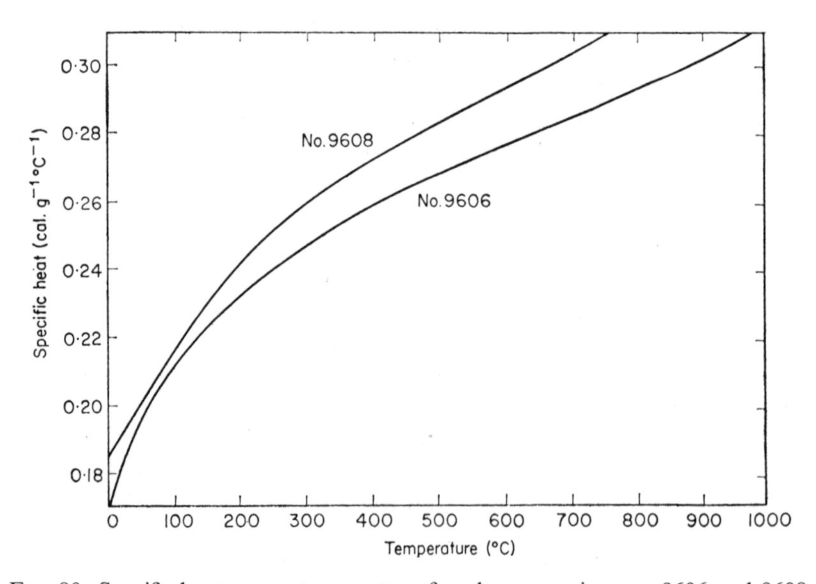

FIG. 80. Specific heat versus temperature for glass-ceramics nos. 9606 and 9608.

TABLE XXX

THERMAL CONDUCTIVITIES OF GLASS-CERAMICS AND OTHER MATERIALS AT 100°C

Material	Thermal conductivity (cal sec^{-1} cm^{-1} °C^{-1})
Glass-ceramics	
Li_2O–ZnO–SiO_2 (c. 5% ZnO)	0·0067
Li_2O–ZnO–SiO_2 (c. 30% ZnO)	0·0052
Li_2O–Al_2O_3–SiO_2 (c. 20% Li_2O)	0·0130
Li_2O–ZnO–Al_2O_3–SiO_2	0·0070
Glasses	
Fused silica	0·0036
Low expansion borosilicate	0·004
Soda-lime-silica	0·0035–0·004
Ceramics	
Pure alumina	0·072
Pure magnesia	0·090
95 per cent aluminia ceramic	0·052
Porcelain	0·004

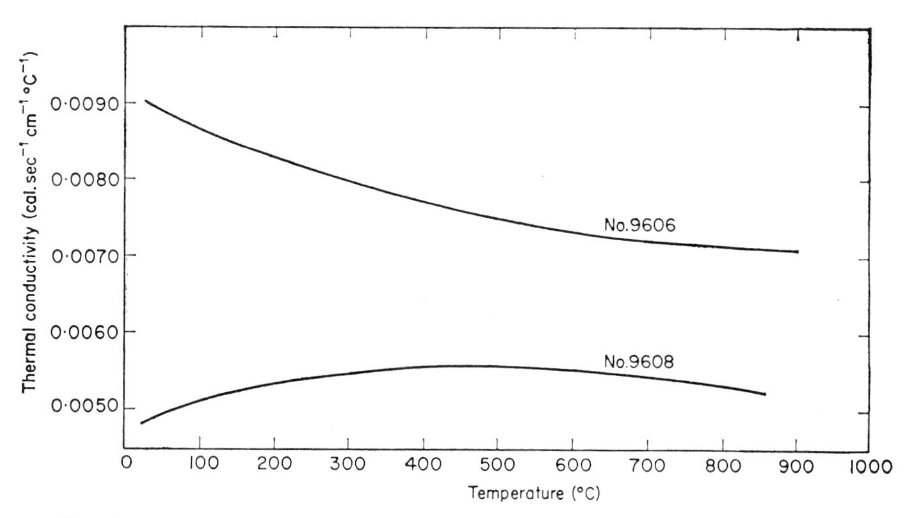

FIG. 81. Dependence of thermal conductivity upon temperature for glass-ceramics nos. 9606 and 9608.

Glass-ceramics have thermal conductivities which are somewhat higher than those of glasses but lower than those of pure oxide ceramics. This is illustrated by the data given in Table XXX.

The effect of temperature on thermal conductivity is shown in Fig. 81 for two glass-ceramics. For glass-ceramic 9606 there is a slight fall of thermal conductivity as the temperature increases and this behaviour is in contrast with that observed for glasses which usually show a slight increase of thermal conductivity as the temperature increases. Crystalline materials, on the other hand, usually show a decrease of thermal conductivity as the temperature increases and the behaviour of glass-ceramic 9606 is consistent with a highly crystalline microstructure containing only a very small amount of a glass phase. The behaviour of glass-ceramic 9608 is unusual, since the thermal conductivity attains a maximum value at a temperature in the region of 450 to 500°C. The overall variation of thermal conductivity throughout the temperature range is slight, however.

b. Thermal diffusivity. In problems involving transient heat flow (where equilibrium has not been established and heat energy must be used in raising the temperature of the body), the quantity thermal diffusivity is employed.

$$\text{Thermal diffusivity } K = k/\rho C_p$$

where k = thermal conductivity; ρ = density; C_p = true specific heat. Figure 82 gives thermal diffusivity data for two commercial glass-ceramics.

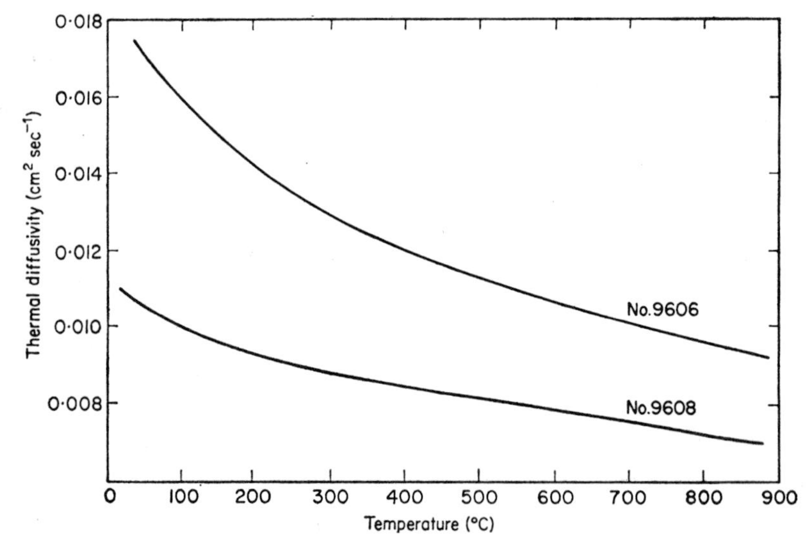

FIG. 82. Thermal diffusivity versus temperature for glass-ceramics nos. 9606 and 9608.

5. *Thermal Endurance*

a. General considerations. If a temperature gradient is established across a material such as a glass-ceramic, internal stresses are generated due to the strain resulting from uneven thermal expansion of the material. For a plate, heated on one face and cooled on the other, compressive stresses are set up on the hotter face and tensile stresses on the cooler face. Under steady state conditions of this type and where the shape of the specimen or external restraints are such as to prevent relief of the stress by bending, the stress at the surface is given by:

$$\sigma = E\alpha\triangle T/2(1-\nu)$$

where σ = surface stress (compressive or tensile)
$\triangle T$ = temperature difference between faces
α = linear thermal expansion coefficient
ν = Poisson's ratio

This assumes a uniform temperature gradient across the thickness of the specimen and it is clear that the actual temperature gradient which can be set up will depend upon the thickness of the material as well as the thermal conductivity.

With sudden heating or cooling, the stresses generated are higher than those for the steady state conditions since initially the temperature gradient is confined to a thin surface layer and the surface stress may attain a value twice that for the steady state conditions. Sudden heating results in the development of surface compressive stresses, but with sudden chilling the stresses are tensile. Failure under tension is much more probable and if the stresses generated exceed the breaking stress of the material, fracture will occur. The extent to which a material can withstand sudden temperature changes without fracture is referred to as its thermal shock resistance and this is often defined in terms of the maximum temperature interval through which the material can be rapidly chilled without fracturing. The resistance to sudden cooling is usually thought to be more important than resistance to heating due to the greater likelihood of failure under the former conditions, but in special cases the resistance to heating may need to be known. Thermal shock testing must be carried out using carefully standardised conditions since the shape of the test specimen can markedly influence the thermal shock resistance. In addition, the nature of the cooling medium used is important since this influences the rate at which heat is extracted from the surface of the specimen. For simplicity and convenience, the thermal shock test often employed comprises quenching specimens from various known temperatures into a fast-flowing water-bath having a standardised temperature.

b. Thermal shock resistance of glass-ceramics. It will be clear that the important factors which determine the thermal shock resistance of a glass-ceramic are the modulus of elasticity, the linear thermal expansion coefficient and the mechanical strength. Poisson's ratio is of secondary importance since the values will not vary greatly from one glass-ceramic to another. Because of their high mechanical strengths compared with normal glasses and ceramics, it is to be expected that glass-ceramics will possess good thermal shock resistances. In addition to this, glass-ceramics having very low thermal expansion coefficients are possible and it is not surprising that these exhibit excellent resistance to thermal shock. Glass-ceramics containing beta-spodumene as a major crystal phase and having thermal expansion coefficients of 5 to 10×10^{-7} can be quenched from temperatures in the region of 700°C into an ice-bath at 0°C without failing. In addition to the requirements for high strength and low thermal expansion coefficient, a low modulus of elasticity favours high thermal shock resistance. It is fortunate that the glass-ceramics having beta-spodumene as the major crystalline phase also have moderately low values of Young's modulus (c. 8·3 to $9{\cdot}0 \times 10^4$ MNm^{-2}). Thus materials of this type possess a valuable combination of properties which confers high thermal shock resistance.

The data given in Table XXXI illustrate the influence of various properties on the thermal shock resistance. The tests were carried out by quenching rod specimens from various temperatures into water at 20°C. Mechanical strength measurements were carried out on sets of quenched rod specimens. The microcracking caused by the thermal shock results in weakening of the material and a reduction of the strength to 90 per cent of the initial value for the unquenched material was taken to indicate that failure by thermal shock

TABLE XXXI

THERMAL SHOCK RESISTANCES OF GLASS-CERAMICS AND SODA-LIME-SILICA GLASS

Material	Temperature interval to cause failure (°C)	Modulus of rupture (lb/in^2)	Modulus of elasticity (lb/in^2)	Thermal expansion coefficient ($°C^{-1}$)
Glass-ceramics				
a	130	25 000	$10{\cdot}6 \times 10^6$	$156{\cdot}9 \times 10^{-7}$
b	160	40 000	$12{\cdot}6 \times 10^6$	$127{\cdot}5 \times 10^{-7}$
c	215	22 000	$12{\cdot}0 \times 10^6$	$64{\cdot}9 \times 10^{-7}$
d	285	28 000	$11{\cdot}2 \times 10^6$	$55{\cdot}5 \times 10^{-7}$
Soda-lime-silica glass	140	16 000	$10{\cdot}0 \times 10^6$	$92{\cdot}0 \times 10^{-7}$

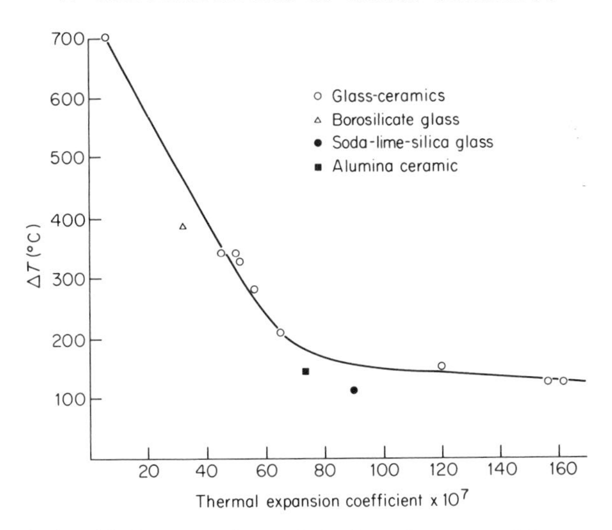

FIG. 83. Relationship between thermal shock resistance and linear coefficient of thermal expansion for glass-ceramics and other materials.

had occurred. It appears that the linear coefficient of expansion is the most important factor determining the relative thermal shock resistances of glass-ceramics. As shown in Fig. 83 when the values of $\triangle T$, the temperature interval causing thermal shock failure are plotted against thermal expansion coefficient the points lie on a smooth curve (expected if thermal shock resistance is inversely proportional to thermal expansion coefficient). It should be noted that the elastic moduli of the glass-ceramics were all fairly similar as were their thermal conductivities. Although the mechanical strengths of the glass-ceramics differed, the variations were insufficient to cause the values of $\triangle T$ to deviate markedly. In Fig. 83, points showing the thermal shock resistances of soda-lime glass and a low expansion borosilicate glass are also given. The values of $\triangle T$ for these two materials both lie below the curve and this is probably due to their significantly lower mechanical strengths as compared with the glass-ceramics. It is somewhat surprising at first sight to note that the value of $\triangle T$ for 95 per cent alumina ceramic is also lower than would be expected from its thermal expansion coefficient. The higher thermal conductivity of this ceramic might be expected to confer a better thermal shock resistance than would be shown by a glass-ceramic of comparable thermal expansion coefficient. That this is not so is likely to be due to the significantly higher elastic modulus of the alumina ceramic.

Quenching into still air has a much less serious effect than quenching into water and this effect has been demonstrated for glass-ceramic *b* in Table XXXI. In tests, in which rods of this material were rapidly removed from a

TABLE XXXII

PROPERTIES OF MATERIALS SUBJECTED TO THERMAL SHOCK BY RAPID HEATING

Material	Thermal expansion coefficient (20–500°C) ($°C^{-1}$)	Modulus of rupture (lb/in²)	Young's modulus (lb/in²)	Thermal conductivity at 100°C (cal sec^{-1} cm^{-1} $°C^{-1}$)
Glass-ceramic	$75·5 \times 10^{-7}$	38 000	$15·1 \times 10^{6}$	0·013
95% Alumina-ceramic	$72·1 \times 10^{-7}$	35 000	$35·0 \times 10^{6}$	0·052

furnace maintained at various temperatures, quenching from a temperature of 700°C did not significantly affect the strength of the glass-ceramic.

There are few data concerning the effects of sudden heating upon glass-ceramics although it is fairly certain that they will be highly resistant to this type of thermal shock. The resistance of a glass-ceramic of the Li_2O–Al_2O_3–SiO_2 type to rapid heating was studied by exposing discs of the material, 5 cm diameter and 0·5 cm thick, to a jet of air which had been preheated to 750°C in a heat exchanger; the air stream had a velocity of 300 feet per second and it was allowed to impinge on one face of the discs for 1 minute. Five glass-ceramic specimens were tested in this way and no failures occurred but in the same test two out of five specimens of a 95 per cent alumina ceramic failed. The thermal expansion coefficients and moduli of rupture of the two materials were similar (Table XXXII) and the higher thermal conductivity of the alumina ceramic would favour the generation of a smaller temperature gradient so that the strain produced would be smaller. On this basis, the better performance of the glass-ceramic is mainly attributed to its smaller modulus of elasticity, which means that for a given strain the stress would be smaller. In addition to this factor, the rate of heat transfer between the air stream and the test specimen may have been different for the two materials and this might also account for their different behaviour.

Chapter 6

APPLICATIONS OF GLASS-CERAMICS

The unique combination of properties that can be achieved for glass-ceramics renders these materials suitable not only for replacing more traditional materials in applications where better cost effectiveness and improved performance are desirable, but also opens up entirely new fields where no alternative material can satisfy the technical demands. Glass-ceramics have achieved wide usage in a number of fields but there are many potential applications still awaiting industrial exploitation. It is the purpose of this chapter to discuss the various actual and potential applications of glass-ceramics in relation to the properties of the materials.

The applications are grouped under major headings describing various areas of engineering and technology.

A. General and Mechanical Engineering Applications

The high mechanical strengths, good dimensional stability and abrasion resistance of glass-ceramics render them suitable for a number of applications in mechanical engineering. Although the mechanical properties may form the main criterion in the selection of glass-ceramics for such applications, in a number of cases thermal and electrical characteristics are also important.

1. *Bearings*

The good mechanical properties of glass-ceramics combined with their ability to take a very smooth surface finish have enabled them to be used for special purpose bearings. Glass-ceramics are superior to conventional ceramics with regard to the surface finish that can be attained since the best surface finish achievable for a 95 per cent alumina ceramic is 200 to 250 nm (8 to 10 micro inches) whereas glass-ceramics can be polished to surface finishes of 12·5 nm (0·5 micro inch). Because of their resistance to corrosion and freedom from oxidation glass-ceramics are suitable for operation under stringent ambient conditions where most metals would suffer unacceptable deterioration. There is also the possibility of operating with minimum lubrication and this is advantageous for operating conditions (eg in space vehicles) where normal lubricants will not function satisfactorily or where normal lubrication techniques cannot be used owing to inaccessibility of the bearing during service. The use of different glass-ceramics or of a glass-ceramic and a metal for the two bearing surfaces, following the well-established principles of using dissimilar metals for rotor and stator components in conventional bearings, is likely to give a minimum wear. Wherever possible, of course, liquid or gas lubrication of the bearing is utilised. The technology exists for the production of various designs of ball, roller or journal bearings in glass-ceramic and for the use of techniques which allow metal shafts and housings to be clad in glass-ceramic.

2. *Pumps, valves and pipes*

The high hardness and excellent abrasion resistance of glass-ceramics suggest their use for the construction of pumps, valves and pipes for handling abrasive slurries. In addition, the good chemical durabilities of many glass-ceramics enable them to be used in contact with corrosive liquids under conditions where many metals would undergo unacceptable deterioration. Stainless steel components can be used in some applications but the superior resistance to wear of glass-ceramics would be advantageous. Pincus (1971) has described the use of a low expansion glass-ceramic containing a beta-spodumene/keatite solid solution for making pump impellors and casings and piping. The low thermal expansion coefficient confers high thermal shock resistance rendering the components suitable for handling hot fluids. The use of techniques which enable metal parts to be firmly attached to the glass-ceramic or the cladding of metal components with an abrasion-resistent coating gives the potentiality for great flexibility of design.

3. *Heat Exchangers*

The need for improved heat exchangers has been underlined by the application of gas turbine engines in land transport vehicles. Without recovery of waste heat, the overall efficiency would be uneconomically low. The favoured design is a regenerative heat exchanger in which the hot exhaust gases first heat the regenerator structure; during the subsequent cycle air is drawn over the heated structure before entering the combustion chamber. This application demands a material of high chemical and dimensional stability and also one having outstanding resistance to thermal shock. Glass-ceramics having near-zero coefficients of thermal expansion fulfil these requirements. The optimum design of heat exchanger consists of a porous disc through which the hot gas and cold air flow in opposite directions. Rotation of the disc through 180° at short intervals ensures that a hot structure is presented to the incoming air on an almost continuous basis. The success of this design depends on efficient sealing between the faces of the disc and the gas and air ducts and this is best achieved using a material which will not change its dimension when it is heated and cooled. That is, the material of the disc should have a coefficient of thermal expansion closely approaching zero. This constitutes another important reason for selecting a glass-ceramic for this application.

The structure of the heat exchanger needs to be such that it will allow relatively unimpeded flow of the air and exhaust gases but will heat up rapidly. The specific heat should clearly be as high as possible to allow maximum heat capacity.

Goss (1962) described a suitable glass-ceramic material having a very low thermal expansion coefficient ($-2 \times 10^{-7}\,°C^{-1}$); this confers dimensional stability and excellent thermal shock resistance. The material also has a high specific heat, is chemically stable and is suitable for long-term operation at temperatures up to 1100°C.

Although heat exchangers can be made by fusing or sintering together thin-walled glass-ceramic tubes, a lighter weight structure can be made by a process described by Hollenbach (1963). In this method a flexible paper carrier is coated with a powdered suspension of a suitable crystallisable glass. The coated paper is then crimped to form corrugated sheets which are assembled with alternate flat sheets to form an open honeycomb-type structure. The assembly is then subjected to a heat-treatment process which burns out the paper and also sinters and crystallises the glass powder to form a strong but thin-walled glass-ceramic structure. The cellular material formed in this manner, can be sawn, drilled, ground and turned with conventional metal working tools.

Additional applications of the cellular glass-ceramic include the manufacture of infrared burners, catalyst supports for chemical engineering use and heat exchangers for cryogenics.

4. *Furnace Construction*

Large furnaces for the melting of glass (known as tank furnaces) are constructed from refractory blocks. Although these, which are often of the fusion-cast variety, are resistant to corrosion by molten glass, problems arise owing to attack at the joints between them where, despite accurate facing of the blocks, penetration by molten glass inevitably occurs. The resultant corrosion not only causes deterioration of the glass quality but also reduces the life of the furnace. The use of a suitable sealant for the joints would obviously be beneficial. The properties required for the sealant include resistance to corrosion by molten glass, stability at the operating temperature of the furnace, reasonably close matching of the thermal expansion coefficient to the material of the block and also a small degree of plasticity in the operating temperature range to allow small movements to occur without rupture of the bond between the sealant and the block.

A glass-ceramic "cement" having these properties has been described by Hayward (1978). This material which is applied in powder or slurry form, partially fuses during the heating of the furnace to wet and bond to the refractory block; it subsequently crystallises *in situ* to produce a glass-ceramic bond. Large scale trials of the glass-ceramic cement have demonstrated its valuable characteristics in significantly reducing corrosion problems.

5. *Other Engineering Applications*

The good wear resistance of glass-ceramics combined with chemical stability have led to their use for jigs and other components required for electro-discharge machining processes. An application making use of the ability of photochemically machinable glass-ceramics is the production of masters from which rubber printing plates can be made. The ability to reproduce fine detail is crucial here but the high strength and durability of the glass-ceramic are also important properties.

Another application, in which the durability of glass-ceramics plays a key role is the production of architectural glass-ceramics. In the United States and Japan use has been made of glass-ceramics for building construction. Products include curtain walling, building blocks and interior finishes for heavy traffic areas such as lifts. Somewhat more exotically, it has been suggested that glass-ceramics might be suitable materials for the construction of undersea structures and for deep submergence vessels. The very high compressive strengths of glass-ceramics together with high chemical stability and absence of corrosion are attractive features for this application.

An important application, which also makes use of the good chemical durability of glass-ceramics is the internal coating of large steel vessels used in the chemical and food processing industries. Vitreous enamel coatings have been used for this purpose for many years but a fine-grained glass-ceramic coating offers advantages. The higher strength of the coatings combined with good resistance to sudden temperature changes allows useful extension of the permissible operating conditions. Sandford *et al.* (1968) have described glass-ceramic compositions for this application which have better acid resistance, improved thermal shock resistance and higher strenghs than the vitreous enamel coatings previously employed.

Glass-ceramics are also finding applications in the important areas of pollution control and new energy sources. One such application is the use of porous glass-ceramic in an afterburner for eliminating hydrocarbons from automobile exhaust systems. Another example concerns the use of glass-ceramic seals in sodium-sulphur batteries. A high temperature sealant, compatible with the beta-alumina ceramic electrolyte is required and the selected material must not deteriorate when in contact with molten sodium. Silica-free glass-ceramics appear to have the necessary properties.

B. Applications in Electrical Engineering and Electronics

1. *Glass-ceramic to Metal Seals*

The joining of metals to insulators plays a very important role in electrical engineering, in microelectronics and in vacuum tube construction. Glass-to-metal seals occupy a very important position in these technologies but where higher temperature processing or operating conditions are required, advantage is taken of the good mechanical properties and refractoriness of ceramics. Glass-ceramics combine many of the valuable properties of glasses and ceramics and their use for sealing to metals for the construction of a range of devices therefore represents a very important application.

The main criterion for selecting a glass-ceramic for sealing to a metal is the achievement of fairly close matching of thermal expansion characteristics to prevent the generation of dangerously high stresses resulting from differential contraction. Depending on the actual application, however, other properties may be important. Clearly a high mechanical strength will be desirable and for high voltage insulators this should be combined with a high dielectric breakdown strength. For vacuum tube applications the ability of the glass-ceramic to withstand a high temperature out-gassing process is needed and if the tube is designed for high frequency applications an additional requirement will be a low dissipation factor.

The versatility of glass-ceramics for metal sealing applications derives mainly from the fact that the thermal expansion coefficients can be varied over an extremely wide range enabling matching of the coefficient to practically every metal. It is not possible to achieve this flexibility with either glasses or conventional ceramics. Figure 84 illustrates some glass-ceramic/metal combinations.

There are two basic processes for sealing glass-ceramics to metals. One of these follows fairly closely the techniques used for joining ceramics (such as alumina) to metals. Glass-ceramic components are formed, usually by pressing, and after the crystallisation heat-treatment the required dimensional tolerances are achieved by diamond grinding. Sealing to the metal components is accomplished in a subsequent operation using an intermediate bond. The latter may be a thin layer of a special sealing glass or glass-ceramic or a layer of a braze metal; this requires the premetallisation of the glass-

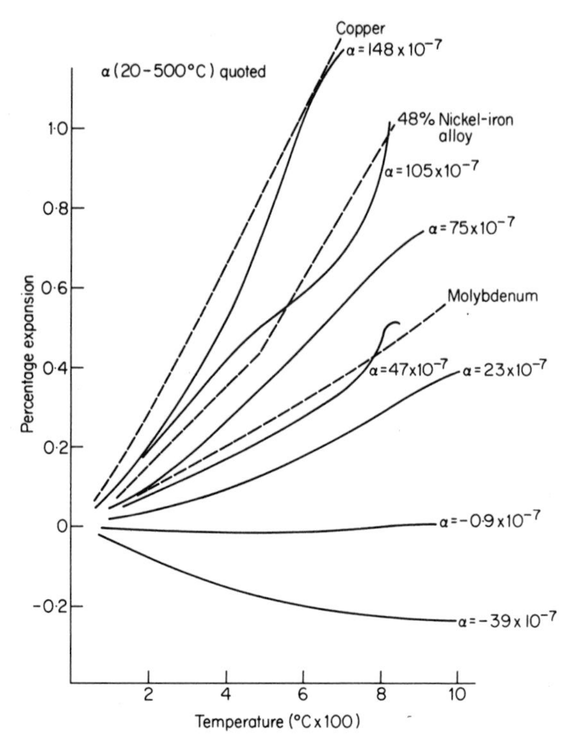

FIG. 84. Thermal expansion curves of metals and glass-ceramics.

ceramic parts or the use of a special braze layer incorporating an "active" metal such as titanium.

An example of the use of intermediate bond sealing was described by Miller and Shepard (1961). A cordierite type glass-ceramic was selected on the basis of its very good dielectric properties and because its thermal expansion closely matched that of tungsten and molybdenum. In addition the gas permeability of the material was low. In the actual tube (described as the "matchbox" tube) two rectangular pressings were sealed edge to edge by a devitrifiable solder glass. Tungsten lead wires and an exhaust tubulation passed through the seal between the two pressings.

Tubes made in this way had excellent characteristics including a shelf life exceeding 5000 hours, capability of continuous operation at 400°C and ability to withstand thermal cycling up to 750°C. They were also resistant to radiation exposure and salt spray tests.

McMillan *et al.* (1966a) described a range of glass-ceramic to metal seals using intermediate bonds. The metals used were copper, nickel and mild steel and the bonding media employed were glasses, glass-ceramics and active metal brazes. It was shown that vacuum tight seals having exceptionally good mechanical strengths could be achieved.

The second basic process for glass-ceramic to metal sealing makes full use of the special characteristics of glass-ceramics since sealing is accomplished while the material is in the glass state and the seal so formed is heat-treated to convert the glass into a high strength glass-ceramic. This process was first described by McMillan and Hodgson (1963a). At first sight there are some apparent difficulties in this process. First, it is only in relatively few cases that the thermal expansion coefficients of the parent glass and the glass-ceramic are the same. Therefore, since the aim is to match the thermal expansion curve of the glass-ceramic to that of the metal, a large mismatch may exist when the parent glass is sealed to the metal. This problem is overcome by ensuring that the seal is not cooled below the strain point of the glass until conversion to the glass-ceramic has been completed. In this way build up of stresses due to differential thermal contraction is avoided. The second possible problem arises because of the volume change that occurs when the glass is converted to glass-ceramic. For metal-sealing glass-ceramics this usually corresponds to about 1 per cent linear contraction. If a strain of this magnitude were not relieved it would generate extremely high stresses and result in disruption of the seal. Fortunately, as has been shown by McMillan *et al.* (1966c) with proper selection of the glass-ceramic composition and heat treatment cycle, the volume change takes place in a temperature range where the glass-ceramic can undergo viscous deformation thus averting the build-up of dangerously high stresses.

There are two basic methods for the direct sealing of glass-ceramic to metal. In the first of these, preheated metal components are positioned in a metal mould, molten glass is introduced and the final seal assembly is shaped by pressing or centrifugal casting. The seal assembly is then inserted into a furnace held at the glass-ceramic nucleation temperature and subjected to a controlled heat-treatment cycle to achieve crystallisation of a glass. The second technique makes use of glass preforms which are assembled together with metal components in graphite moulds. The assemblies are subjected to a heat-treatment cycle in a controlled atmosphere (to prevent oxidation of the metal parts and deterioration of the graphite moulds). The heat-treatment causes the glass to seal to the metal and subsequently to crystallise to give a glass-ceramic closely matched in thermal expansion to the metal.

One advantage of the first sealing process is that it eliminates the need for preshaping the glass-ceramic parts to precise tolerances such as is necessary when a separate sealing operation is employed. It is possible to seal a number of metals to suitable glass-ceramics by this method; the metals include various nickel-iron alloys, chrome-iron alloys, a number of steels (including inexpensive mild steel) and copper. The ability to seal ceramic-type materials to mild steel or copper is especially valuable because the former metal is widely used as a structural material in the electrical engineering industry and the latter is the most commonly used conductor.

Figure 85 illustrates the basic designs of seal assembly that can be made by the direct sealing process. The application of the technique for the manufacture of vacuum tube envelopes is shown in Fig. 86 and Fig. 87 illustrates various lead through seals and multi-pin substrates for electrical applications.

The electrical and other important properties of two glass-ceramics suitable for metal sealing are given in Table XXXIII. Glass-ceramic A, designed for sealing to mild steel has a high dielectric breakdown strength and is thus suitable for the manufacture of high voltage insulators. Glass-ceramic B, which is sealed to copper, has low dielectric loss characteristics and is therefore suitable for applications in electronics.

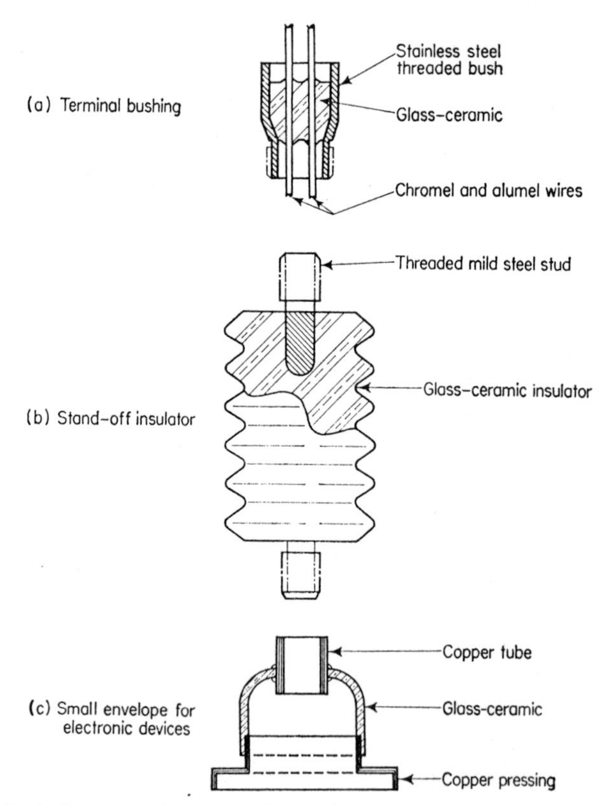

FIG. 85. Seals between glass-ceramics and metals made by the direct moulding process.

FIG. 86. Glass-ceramic/metal envelopes for vacuum switches. (This figure and Fig. 87 are reproduced by kind permission of the Materials Division, G.E.C. Power Engineering, Stafford Laboratory.)

FIG. 87. Glass-ceramic/metal lead-through seals and substrates made by the direct sealing process.

TABLE XXXIII

PROPERTIES OF GLASS-CERAMICS DEVELOPED FOR SEALING TO MILD STEEL (A) AND COPPER (B)

Property and units	A	B
Density (g cm^{-3})	2·52	3·14
Modulus of rupture (MNm^{-2})	286	251
Young's modulus (GNm^{-2})	92	82
Thermal expansion coefficient 20 to 600°C, (°C^{-1})	140×10^{-7}	178×10^{-7}
Deformation temperature (°C)	850	750
Dielectric breakdown strength ($kVmm^{-1}$)	447	30
Volume resistivity at 20°C (ohm cm)	$3{\cdot}2 \times 10^{15}$	$1{\cdot}9 \times 10^{15}$
Permittivity (20°C) at:		
1 MHz	5·0	5·8
10 MHz	—	5·8
100 MHz	5·3	5·8
1000 MHz	5·5	5·9
Loss angle, tan δ (20°C) at:		
1 MHz	0·0023	0·00032
10 MHz	—	0.00028
100 MHz	0·00050	0·00020
1000 MHz	0·00081	0·00053

2. *Insulators and Bushings*

The high dielectric breakdown strengths and mechanical strengths of glass-ceramics offer advantages for insulators as compared with conventional electrical porcelains because thinner sections can be used resulting in weight savings and greater freedom of design. The smooth surface which does not require glazing is beneficial in alleviating the build-up of surface pollutants which can cause electrical breakdown.

Very often, the insulator component must be hermetically sealed to a metal, as for example in transformer bushings, and the ability to match the thermal expansions of glass-ceramics to suitable metals enables joints of greater reliability to be realised since stresses generated during thermal cycling will be low. Also, there is the possibility, as discussed earlier, of sealing the glass-ceramic to the metal giving joints of great integrity.

3. *High Temperature Electrical Insulation*

An application of glass-ceramics which makes use of their high electrical resistivity is the insulation of wires and other conductors for use at high temperatures, eg for thermocouple and instrument cables in atomic reactors where organic types of insulation would be quite unsuitable. McMillan (1970) has discussed special types of glass-ceramics developed for this application; these are phosphate rather than silicate compositions and they are free from alkali metal oxides. A typical glass-ceramic of this type has a high volume resistivity (10^{12} ohm cm at 300°C and 10^{9} ohm/cm at 500°C). The glass-ceramic can be applied to wires of various compositions by a continuous coating process. Adherent coatings are formed on chromel and alumel thermocouple wires and on other wires including platinum and platinum alloys. In addition to high electrical resistivity, the glass-ceramic has excellent dielectric breakdown strength. For twisted pairs of wires provided with a glass-ceramic coating 25 nm in thickness the breakdown voltage is 600 to 800 V. The glass-ceramic insulation can be safely used at temperatures as high as 700°C without deterioration.

4. *Preformed Circuitry for Electronics Applications*

The electronics industry makes considerable use of preformed circuitry of various types ranging from printed circuits to other more sophisticated technologies. The properties of glass-ceramics commend them for these types of application. In particular, the good electrical characteristics and dimensional stability at elevated temperatures offer distinct advantages.

Glass-ceramic printed circuit boards have been made from chemically machined photosensitive glass-ceramic. This allows boards containing a precisely spaced pattern of fine holes to be achieved. In another form of printed circuit board, metal foils which may be of copper, silver or nickel are bonded to glass-ceramic sheets in a direct-sealing process (McMillan and Hodgson, 1961). The foils can be etched in standard processes to produce conductor patterns. Silver or nickel foils are preferred for high temperature applications since these metals are resistant to oxidation.

A novel process for producing surface-metallised glass-ceramic was originated by McMillan and Hodgson (1963b). This is based on the discovery that if glass-ceramics containing 0·5 to 7·5 weight per cent cupric oxide are heat treated in a reducing atmosphere, an adherent surface film of metallic copper is produced. This process involves migration of copper ions from the interior of the glass-ceramic to the surface where they are reduced to metallic copper resulting in the deposition of a metal film of significant thickness. The copper film can be etched by standard methods to produce conducting paths. Since the metal layer is produced on all exposed surfaces, metal coated tubes and other more complex shapes can be produced which have potential applications in a variety of electronic devices. In fact Seki (1974) has followed up these possibilities and developed techniques to produce printed circuit boards, resistors and components for electron microscopes.

Byer (1962) described an interesting technique for the production of highly stable circuits. The process is based on the observation that if pieces of glass-ceramic are placed in contact and heated, interdiffusion takes place and the parts are transformed into a monolithic body. Bonding can either be accomplished during the normal crystallisation cycle or as a subsequent operation.

In the production of circuits (described as submerged circuits) plates of a suitable glass-ceramic are photochemically machined, using the processes described in chapter 4, to produce a desired pattern of grooves, slots or holes. When the plates are bonded together the conductor pattern is defined by the configuration of the slots, etc. The conductors may be of a relatively refractory metal which will not melt at the bonding temperature (usually about 850°C), in this case the conductors are inserted or deposited in the conductor channels before heat-treatment. Alternatively, the conductors may be of a low melting point metal or alloy and in this case the molten metal is drawn by suction into the passageways of the glass-ceramic article after the bonding process.

This process offers the possibility of producing multilayer circuits for microelectronics that are completely isolated from external factors such as humidity, pressure and temperature effects. Organic plastics are less satisfactory then glass-ceramics for this application because they are not fully impermeable to moisture and thus deterioration of insulation resistance can occur.

5. *Substrates for Microelectronics*

In the production of computers and similar equipment it is necessary to mount the silicon chips each containing a large number of devices made by highly sophisticated diffusion processes, onto an electrically insulating substrate and to provide a means of interconnecting the various active components.

Alumina ceramic has been extensively used in this application but glass-ceramics can offer some advantages. These will become apparent if the processes for depositing conductor patterns are considered. In one of these, known as the thick film technique, a paste containing the conducting metal (usually gold) is applied in the required configuration by screen printing. In the second method, the conductor pattern is deposited by vacuum evaporation technique using masks to define the configuration of conductors. The ease with which glass-ceramics will form firm bonds to a variety of metals offers advantages in thick film technology and the very smooth surface finish that can be achieved is a great advantage for thin film techniques. In many designs of substrates, it is necessary to incorporate firmly bonded connector pins which provide a plug-in capability. This is readily achieved with glass-ceramics using the direct-sealing technique described earlier and substrates containing precisely defined arrays of sealing alloy pins have been produced.

One possible disadvantage of glass-ceramic as a substrate material is that its thermal conductivity is considerably lower than that of alumina ceramic. In some semiconductor devices the power dissipation per unit area is quite high and the heat must be removed to prevent overheating and consequent damage to the device. To solve this problem, McMillan *et al.* (1970) proposed to utilise a copper plate of substantial thickness to act as the ground plane and as a heat sink for a composite substrate. In this device, glass-ceramic coatings are applied to opposite faces of the copper plate (about 3 mm in thickness). Connections are provided between the two faces by copper wires passing through holes in the plate and sealed by means of an insulating glass-ceramic. Thick film circuitry is applied to each face of the plate and this can be overlayed by further glass-ceramic coatings and additional circuit layers to give a multilayer capability. A region adjacent to one edge of the plate is left uncoated for the application of forced cooling. This type of substrate has been shown to have very good thermal dissipation characteristics enabling close spacing of devices; this is important in enabling computers capable of operating at very high speeds to be built.

6. *Capacitors*

Mention has already been made in chapter 5 of special glass-ceramics containing ferro-electric crystal phases and having high permittivities. While it is true that high permittivity ceramics can be produced by conventional techniques, the use of the glass-ceramic process offers certain advantages. One of the principal advantages is that the glass-ceramic composition, while in its glass state, can be drawn into a very thin film by a continuous process, whereas the manufacture of very thin plates of conventional ceramics is extremely difficult. Thus the production of dielectric layers giving high capacitance per unit volume is achieved more easily with glass-ceramics than with previous materials.

The process for making a glass-ceramic capacitor comprises stacking layers of the thin glass sheet and of a conducting metal, heating the assembly to soften the glass and to fuse the edges of the glass laminations together and afterwards heat-treating the assembly to crystallise the ferro-electric compounds. The relatively low dielectric losses, high dielectric breakdown strengths and good insulation resistance of suitable glass-ceramics of this type are valuable additional characteristics for this application.

7. *Sealing and Bonding Media*

There are many instances in the manufacture of electronics components where a sealing medium is required to join together components of another material without subjecting these components to temperatures that will cause melting, distortion or other structural degradation. Specially formulated glass-ceramics have been derived for such applications and bonding together of glass or ceramic parts, metals and semiconductors can be accomplished.

The use of solder glasses (usually lead-zinc-borate types) to bond glass and metal components together has been known for some years. In this technique a thin layer of the solder glass flows and wets the sealing faces at temperatures where the main components of the device are still rigid. Thus distortion of the glass parts is avoided. Solder glass seals, however, suffer from the limitation that the subsequent processing and operating temperatures must be restricted to avoid softening the solder glass. Conversion of the solder glass to a glass-ceramic would permit higher processing temperatures and this is advantageous because the attainment of high vacuum is dependent upon being able to outgas the tube at the highest possible temperature.

Solder glass-ceramics have been described by Claypoole (1959) who employed lead-zinc-borate materials having approximate weight percentage composition: PbO: 70–80; ZnO: 10–15; B_2O_3: 6·5–10 plus minor constituents such as Al_2O_3 and SiO_2. The finely powdered glasses are suspended in an organic vehicle and applied as a thin film to the sealing faces. The precoated faces are placed in contact and heated to a temperature of 420 to 450°C for 30 minutes; the glass flows to accomplish sealing but it also devitrifies so that it becomes more refractory. Seals of this type can be heated to within 20°C of the sealing temperature.

The lead-zinc-borate compositions are suitable for sealing together glasses having thermal expansion coefficients of $80–120 \times 10^{-7}\,°C^{-1}$ and an important commercial application is in the sealing of faceplates to cones in colour television tube manufacture.

Other compositions of the zinc borosilicate type are available for sealing together glasses or other materials having expansion coefficients in the range $30–50 \times 10^{-7}\,°C^{-1}$. The weight percentage compositions of these glass-ceramics are: ZnO: 60–65; B_2O_3: 20–25; SiO_2: 10–15 plus certain minor constituents.

There are applications requiring the coating of semiconductor materials, such as silicon, or their bonding together or to substrates of other materials. To avoid structural damage to the silicon which would cause unacceptable deterioration of the semiconducting properties, the bonding temperature must be well below the melting point (1410°C); also stresses caused by differential contraction must be low otherwise structural defects are generated. It is desirable, on the other hand, that the bonding medium should be sufficiently refractory to allow diffusion processes and other operations to be carried out at temperatures in the region of 1200°C. Glass-ceramics of the $ZnO–Al_2O_3–SiO_2$ type to which additions of CaO, BaO or B_2O_3 are made have been shown by McMillan and Partridge (1969) to be suitable for this application. Glass-ceramics having thermal expansion coefficients between 30×10^{-7} and 40×10^{-7} °C^{-1} were considered to be suitable for bonding to silicon. These compositions contain zinc orthosilicate ($Zn_2 SiO_4$) and zinc aluminate ($ZnAl_2O_4$) as major crystalline phases. The electrical resistivities of these glass-ceramics are high, being generally greater than 10^8 ohm cm at 500°C.

For the coating of n-type silicon, the glass was ground to a fine powder and applied in the form of a slurry by a flow-coating technique. It was found to be necessary to pre-oxidise the silicon to give an oxide layer of 0·5 μm thickness to prevent diffusion of glass-ceramic constituents into the silicon. By doing this, and restricting the firing temperature for fusion of the glass-ceramic to 1180°C, the extent of damage to the underlying silicon was very slight. Coating thicknesses between 10 and 75 μm could be applied and these were refractory at temperatures up to 1100°C. It was also shown that the glass-ceramic coatings could be used to bond together silicon components.

A problem exists in sealing together components of low expansion borosilicate glass at a sufficiently low temperature to prevent distortion. For example, in certain designs of television camera tube it is required to seal an optically flat face-plate to the main tube without causing deterioration of the optical characteristics. McMillan and Partridge (1970) developed composite sealing media in which a glass-ceramic having a negative thermal expansion coefficient was combined with a low fusion point $PbO–ZnO–B_2O_3$ glass. By this means composite materials were derived having thermal expansion coefficients of 30×10^{-7} °C^{-1} or less but which also could be used for relatively low temperature sealing to avoid distortion of optical components. In a later development the same investigators (McMillan and Partridge, 1973) were able to reduce the softening points of the matrix glass by the inclusion of halogens thus enabling composite sealing media with even lower flow points to be derived.

8. *Electro-optical Materials*

High permittivity glass-ceramics containing ferroelectric crystals can be produced in forms transparent in the visible spectrum provided the heat treatment is controlled to ensure that the crystals are small enough. To achieve this condition, the crystal size has to be an order of magnitude less than the wavelength of visible light. Suitable materials include those with sodium niobate as the ferroelectric phase (Borelli, 1967 and Borelli and Layton, 1969) and those containing lead titanate (Kokubo and Tashiro, 1973). These glass-ceramics exhibit an interesting electro-optical effect in that the plane of polarisation of a beam of polarised light passed through a block of the material is rotated when an electric field oriented at 45° to the plane of polarisation is applied. The magnitude of the effect is proportional to square of the electric field. A similar type of electro-optical behaviour is found for some ferroelectric single crystals in the para-electric state but Borelli (*loc. cit*) points out that in crystallised glass the activity observed must be due, at least to some extent, to the ferroelectric state and to phenomena related to reorientation of domains under the applied electric field.

C. Lighting and Optical Applications

1. *Lamp Envelopes*

High intensity light sources such as discharge lamps require materials that possess a number of properties in addition to transparency or translucency. High mechanical strength is desirable and since the efficiency of the lamps increases as the operating temperature is raised, ability of the material to withstand high temperatures without deformation is also advantageous. For these reasons, translucent alumina ceramics have been adopted for discharge lamp manufacture because of their superiority to glasses. A drawback of such ceramics, however, is that it is relatively difficult to seal them to metal electrodes.

It is therefore of interest to consider whether transparent glass-ceramics would be suitable for this application since the sealing of these to metals presents fewer problems. The majority of transparent glass-ceramics have very low or near zero thermal expansion coefficients and therefore problems of thermal expansion mismatch with suitable metals such as tungsten or molybdenum exist. The use of a graded glass-ceramic seal would enable this difficulty to be overcome but may be uneconomic. It has been found possible, however, to produce refractory, transparent glass-ceramics closely matched in thermal expansion coefficients to molybdenum and tungsten and capable of being sealed to these metals. Stryjak and McMillan (1978) have described glass-ceramics of this type in which the major crystalline phase is gahnite.

While glass-ceramics based on silicate glasses have good resistance to halogen vapours used in some types of discharge lamps, they undergo reaction with sodium vapour resulting in colouration and darkening of the material. Hence these glass-ceramics are unsuitable as envelope materials for sodium vapour discharge lamps. Silica-free glass-ceramics can be produced that are resistant to sodium vapour and Plesslinger *et al.* (1975) have described modified calcium aluminoborate compositions of this type. These glass-ceramics are stated to be suitable either for manufacturing the complete lamp envelope or as sealing media for bonding a sodium vapour-resistant ceramic envelope to suitable metal electrodes.

2. *Components for Lasers*

Stable performance of gas lasers requires dimensional stability of components used in their construction. For this reason, the use of zero expansion glass-ceramics for the production of envelopes and other components is of interest. For the envelope, another requirement is low permeability to the filling gas, especially helium, and glass-ceramics are generally superior in this respect to conventional low expansion glasses.

3. *Telescope Mirrors*

The performance of large astronomical telescopes of the reflecting type is critically dependent on the properties of the material used for the mirror blank. This material must of course be capable of taking an optical polish and in addition the lowest possible coefficient of thermal expansion is desirable to minimise distortion of the mirror as a result of unavoidable temperature gradients. Formerly, large telescope mirrors were made of low expansion ($30 \times 10^{-7}\,°C^{-1}$) borosilicate glass. The use of fused silica glass with a much lower expansion coefficient ($5 \times 10^{-7}\,°C^{-1}$) though it results in great improvements in dimensial stability, creates severe technological problems in fabrication of the mirror blanks because of the refractory nature of this glass. The advent of zero-expansion glass-ceramics which can be optically polished has allowed a very significant advance to be made in the performance of large reflecting telescopes.

Mirror blanks of this type have been manufactured both in the United States and in Germany. Blanks up to 4 metres in diameter, weighing about 18 000 kg have been made by a casting process followed by a controlled heat-treatment cycle. It is considered feasible that blanks up to 10 metres in diameter could be made by the same process.

D. Applications in Aerospace Engineering

The development of high-speed aircraft and space craft has created demands for materials capable of operating under extremely stringent conditions. It is likely that glass-ceramics can meet a number of the challenges that are being posed and these materials are likely to play an increasingly important role in this field.

1. *Radomes*

In aircraft and missiles flying at high speeds it is necessary to provide a protective enclosure known as a radome for the radar aerials. Missile radomes usually form the nose of the missile and can be hemispherical, conical or ogival in shape. Aircraft radomes may be carried above or below the fuselage and are of suitable aerodynamic shape. Glass-fibre reinforced resin is satisfactory for some types of radome but at very high speeds degradation owing to friction heating by the atmosphere is a problem. Ceramic type materials do not suffer from this disadvantage and have therefore been adopted for the manufacture of certain designs of radomes.

The requirements for an acceptable radome material for application in high speed flight can be summarised:

(*i*) homogeneity and uniformity of dielectric constant to ensure minimum distortion of the radar signals by the radome

(*ii*) a low temperature coefficient of dielectric constant

(*iii*) low dielectric losses at the operating frequency (eg 10 000 MHz)

(*iv*) a high mechanical strength which is maintained at high temperatures

(*v*) good thermal shock resistance since a missile re-entering the dense layers of the atmosphere undergoes rapid heating

(*vi*) good rain erosion resistance. A high speed aircraft or missile passing through a rainstorm is subjected to very severe impacts as a result of collision with raindrops in the line of flight.

High alumina ceramics possess many of the foregoing physical characteristics. However, since the geometry of the radome is critical both with regard to contours and wall thickness there are difficulties in producing alumina ceramic radomes with the necessary precisely controlled dimensions. The large shrinkages during firing often accompanied by distortion mean that extensive grinding is necessary to attain the required shape.

The use of glass-ceramics for radome production alleviates the problem of achieving the precise shape since the dimensional change accompanying the crystallisation process is small. This together with the exceptional uniformity of dielectric properties enables the final correction of the radome shape by grinding to be minimised.

Glass-ceramics suitable for radome manufacture are of the alkali-free aluminosilicate type. One example is a glass-ceramic containing cordierite as the major crystalline phase and a material of this type has been extensively used in the United States. This glass-ceramic is prepared from a titania nucleated MgO–Al_2O_3–SiO_2 glass which is melted in platinum-lined tank furnaces to ensure exceptional homogeneity and reproducibility.

2. *Infrared Transparencies*

Materials having high strengths, good rain erosion resistance and high transparency in various regions of the infrared spectrum are required for infrared-activated guidance systems and for guidance and ranging equipment using laser beams. Optically transparent glass-ceramics were described in chapter 4 but it should be added that some glass-ceramics which appear opaque in the visible region, are transparent in the infrared region. If the crystals present are smaller then the wavelength of light under consideration, light scattering does not occur and the glass-ceramics are transparent to that light. Since many glass-ceramics contain crystals less than 1 μm in size, infrared radiation of longer wavelengths can pass through them without undergoing scattering. For glass-ceramics containing silica, however, the transmission is limited to a wavelength of about 4·5 μm owing to the occurrence of the first overtone of a strong fundamental absorption band of the Si–O bond. Thus, high strength glass-ceramics possessing useful infrared transmissions in the wavelength range 1 to 4·5 μm are possible and these could be valuable for some applications.

In cases where transparency is required at higher wavelengths, it is necessary to employ materials free from silica or other oxides that give rise to absorption at the shorter wavelengths. The possibility has been explored of crystallising non-oxide glasses of the chalcogenide type with the aim of combining the excellent infrared transparencies of these glasses with improved strength and refractoriness. Unfortunately, some of the crystals precipitated from chalcogenide compositions are electronic semiconductors having band gaps of the order of 1 eV. This, of course, gives rise to an absorption edge close to 1 μm and the materials are opaque to wavelengths greater than this.

3. *Application of Glass-ceramics as Thermal Protection*

The low densities of glass-ceramics, combined with good mechanical properties suggest that they might be useful for critical parts of aircraft structures. One such application could be in leading edges of wings for high speed aircraft where frictional heating effects are high. The ability to coat metal components with smooth, adherent glass-ceramic coatings also creates valuable possibilities and the prospect of reinforcing glass-ceramics with metal fibres to produce lightweight, tough composites also offers prospects of novel materials for use in air and space craft.

E. Applications in Nuclear Engineering

Engineering for nuclear power production has meant that many new materials have had to be developed and applied since orthodox materials are unable to meet some of the searching conditions of temperature, pressure and radiation flux. Some potential applications for glass-ceramics have already emerged but the need to conserve critical raw materials could render glass-ceramics more attractive in the future since the raw materials used for glass-ceramics are, for the most part, abundant and comparatively inexpensive.

1. *Reactor Control Rod Materials*

A number of materials can be used for reactor control rods, including boron-containing steels. In these, however, only a limited proportion of boron can be incorporated so that such control rods are bulky and heavy. High concentrations of boron and of other elements having high capture cross-sections for thermal neutrons, such as cadmium and indium, can be incorporated in glass-ceramics. McMillan and Hodgson (1959) described two systems of glass-ceramics of this type. One group of materials contains silica as a glass-former and the other boric oxide. Both types of material contain high proportions of cadmium oxide though the amount is usually lower in the borate compositions since boric oxide itself has a high neutron-capture cross section. Indium oxide is usually included also to give absorption of thermal neutrons over a wider range of energies. A typical boron-containing material has the weight percentage composition: CdO: 30; In_2O_3: 5; B_2O_3: 35; K_2O: 15; TiO_2: 15 and a boron-free material CdO: 60; In_2O_3: 5; SiO_2: 25; Nu_2O: 5; CaF_2: 5; Au 0·027.

The glass-ceramics derived from the two systems have satisfactory mechanical strengths. For the borate type materials the thermal expansion coefficients are in the range 105 to $148 \times 10^{-7}\,°C^{-1}$ (20–500°C) and deformation temperatures range from 750°C to 850°C. Corresponding data for the silica based materials are 65 to $100 \times 10^{-7}\,°C^{-1}$ and 850° to 1000°C.

Encapsulation of the glass-ceramics in stainless steel is necessary for reactor applications to prevent "poisoning" of the reactor if accidental fracture of the glass-ceramic occurs. The relatively high thermal expansions of the glass-ceramics are advantageous since this allows them to maintain good thermal contact with the stainless steel sheath over a range of temperatures. This is essential to allow efficient removal of heat generated in the glass-ceramics owing to capture of neutrons. In addition to their use in control rods, the neutron-absorbing glass-ceramics may be useful as biological shielding materials.

2. *Seals for Reactor Use*

In the operation of nuclear power plants it is necessary to monitor the temperature and other important parameters at a number of points within the reactor. Thus large numbers of thermocouple and instrument cables need to be led out of the reactor, often through restricted passages. The cables used are generally sheathed in stainless steel and are of the mineral-insulated type employing compressed magnesia powder as the insulant.

It is necessary to seal the ends of the cables to prevent ingress of moisture which would drastically reduce the insulation resistance. Although brazed ceramic seals have been employed for this purpose, there are severe difficulties in making these of a sufficiently small size. The use of glass-ceramics in a direct sealing process has enabled twin-lead seals with overall diameters as small as 2·5 mm to be made. The seals, which are brazed to the ends of the cables, are vacuum-tight and are not adversely affected by subjection to gas pressures of 21 kg cm^{-2} (300 lb in^{-2}) or by thermal cycling at 20°C per minute between normal ambient temperature and 700°C.

3. *Disposal of Radioactive Wastes*

The processing of spent fuel rods from fission reactors generates large quantities of waste containing radionuclides some of which have long half lives. The adoption and widespread use of the fast breeder reactor, if it takes place, will result in a large increase of the quantities of highly dangerous radioactive wastes.

At the present time the wastes are stored in the form of solutions and clearly this is very unsatisfactory in view of the difficulties of ensuring that no leaks into the environment ever occur during the necessarily very long storage period. For this reason, the nuclear power industry is anxious to develop means of solidifying the wastes to achieve a stable and more manageable form. Of several possibilities, the one that seems most promising and which is the subject of intensive research and development throughout the world, is to combine the radioactive wastes with suitable oxides (eg SiO_2 and B_2O_3) and to convert them into a chemically stable glass. Blocks of this material could then be suitably encapsulated and stored under controlled conditions with safety.

It seems likely that glass-ceramics technology could play a valuable role in this development. One requirement for the storage medium is that it shall have very good chemical stability to prevent the leaking out of radioactive constituents (eg caesium isotopes) if accidental contact with water occurred. Physical stability and high strength are also desirable. In addition, since the radioactive decay of isotopes contained in the material will generate heat, it is necessary to have good temperature stability combined with as high as possible thermal conductivity to allow adequate heat removal from the blocks. In all of these respects, it is probable that controlled crystallisation of the storage glass to convert it to a glass-ceramic would be of benefit.

F. Application in Medical and Related Fields

The general chemical inertness of glass-ceramics combined with high mechanical strength and other good physical properties has generated interest in possible applications in the biomedical field.

MacCulloch (1968) has described the production of artificial teeth from Li_2O–ZnO–SiO_2 glass-ceramics. In a related application, glass-ceramic powder is used as a filler in polymeric composite dental restorative materials. The hardness and good abrasion resistance of the glass-ceramic filler confer advantages in this application. A further benefit is that by using a glass-ceramic of very low or even negative thermal expansion coefficient, a composite material can be produced having a thermal expansion coefficient closely matched to that of the natural tooth material. This is advantageous in helping to preserve the integrity of the bond between the tooth and the filling. Müller (1974) developed special glass-ceramics having high absorption for diagnostic X-rays for use in composite dental restorative materials. This represents a valuable aid to dental surgeons in facilitating examination.

There is considerable interest in the possibility of replacing damaged or diseased bones with suitable implants. The materials used must have high strengths and at the same time must be non-toxic and compatible with the environment within the body. Special metal alloys have been used and in some cases polymeric materials. For some applications high resistance to wear is necessary, for example in artificial hip joints used in the treatment of patients suffering from severe arthritic conditions. It has become clear that materials possessing a better combination of properties than those utilised so far would be beneficial.

Glass-ceramics could be of value in this application because of their high strength, good abrasion resistance and chemical inertness. Furthermore, it seems possible that materials possessing a valuable combination of chemical and physical characteristics could be developed. Glass-ceramics based on phosphate glasses might be more suitable than those derived from silicate systems and also the possibility exists of producing microporous materials. In this way it is conceivable that glass-ceramic implants might be derived that would allow growth of the natural bone structure into and around them and clearly this would be desirable.

Chapter 7

THE RESEARCH AND DEVELOPMENT POTENTIAL OF GLASS-CERAMICS

Significant progress has been made in the science and technology of glass-ceramics during the past two decades. Nevertheless, there are still a number of areas where a clearer understanding of fundamental processes is necessary. Achievement of this is likely to be important to Material Science in general because many of the processes underlying the controlled crystallisation of glasses and the relationships between glass-ceramic microstructures and physical properties are identical with or closely similar to those for other important classes of materials. In addition, the current awareness of the need to conserve energy and raw materials has highlighted the desirability of using materials in new and more effective ways. Glass-ceramics with their outstanding flexibility of composition and properties are likely to assume an important place in the development of new technologies designed to make better use of the world's resources.

It is believed, therefore, that a brief examination of the future role of glass-ceramics as a topic for basic research and as technologically important materials is worthwhile.

A. Glass-ceramics Research

Crystal nucleation and growth processes are crucial for the development of glass-ceramic microstructures from initially amorphous glasses. While it is true that much progress has been made in recent years on these topics, there are still many unanswered questions and a continuing need exists for more theoretical and experimental studies.

While the basic ideas concerning sub-liquidus immiscibility and the development of two phase glass structures either by a nucleation and growth process or by spinodal decomposition have become well established, the effects of prior phase separation upon crystal nucleation and growth are less well understood. This is clearly a field for further research since in many glass-ceramics the initial occurrence of glass-in-glass phase separation seems to be an essential step in the development of the desired microcrystalline structure. It is known that crystal nucleation and growth processes are both influenced by prior phase separation. Generally, the nucleation density is increased and crystal growth rates are reduced; both of these effects favour the development of fine-grained microstructures. The detailed influence of phase separation upon diffusion processes has not been established, however, and this would appear to be essential if the role of phase separation in subsequent crystallisation is to be understood. A further point of interest concerns the possible differences in crystallisation behaviour for glasses which undergo spinodal decomposition as compared with those that phase separate by nucleation and growth. The marked differences between the two kinds of phase-separation with regard to the composition gradients at the interfaces between the two phases would be expected to lead to different behaviour. It should be borne in mind, however, that the phase interfaces in spinodally decomposed glasses become sharp after prolonged heat-treatment and therefore any effects might only be observable during the initial stages of crystallisation.

There is some evidence for "memory" effects in certain glasses; by this is meant that glass which has been crystallised and subsequently remelted tends to crystallise more readily during subsequent heat-treatment. Also for some compositions, the melting process and the rate of cooling of the glass through temperature ranges well above the nucleation range appear to influence the subsequent crystallisation. It would be of great interest to determine whether such effects imply the existence in the glass melt of small regions possessing a higher degree of order than would normally be expected for the liquid state. Such regions might represent remnants of the crystalline structures of materials used to prepare the glass or of structures produced by previous crystallisation of the glass. Detection of such regions, if they exist, will depend upon the development of techniques for establishing the atomic arrangement in glasses at distances beyond the first co-ordination shell. This is a difficult problem but progress is likely using techniques such as EXAFS (Extended X-ray Absorption Fine Structure) studies and Raman Spectroscopy. The latter has already given evidence of the possible existence in borate glasses of "molecular" groupings of greater complexity than would be postulated to exist by classical random network theory of glass structure. Thus development of ideas concerning the structure of glasses and melts may result in a re-examination and revision of ideas concerning glass crystallisation.

It is evident that the minor constituents of glass-ceramics play an extremely important role in modifying the course of nucleation and crystallisation. Foremost among these are those constituents added as nucleating agents and while partial understanding of the role of some of these has developed during recent years, it would be untrue to claim that all questions have been answered. There is still scope for considerable research upon oxide nucleation catalysts to determine fully the relative importance of effects upon sub-liquidus immiscibility, effects upon interfacial energy between embryo crystals and the glass phase and the possibility of the nucleant either alone, or in combination with another oxide, precipitating out to provide heterogeneous nuclei for the epitaxial growth of other crystal phases. A very intriguing question concerns the observation that two nucleating agents exhibit a synergistic effect in that the combination is more potent in promoting nucleation than either nucleant alone.

Minor constituents, other than those generally regarded as nucleating agents, can exert a significant effect upon the glass-ceramic microstructure. For example, the observation that small additions of the larger alkali metal ions to glass-ceramics basically of the lithia-silica system, promote the formation of quartz rather than the undesirable cristobalite is of great practical importance and deserves investigation. Another minor constituent of glasses and glass-ceramics, the presence of which is often unsuspected, is water in the form of hydroxyl ions bonded into the glass structure. Water in this form is very difficult to eliminate from molten glass except by vacuum melting and its effect upon the glass crystallisation process could be profound. One effect of the presence of hydroxyl ions is to lower the viscosity of glass; thus the diffusion processes involved in crystal nucleation and growth can take place more readily. Water has long been recognised as a "mineraliser" and a detailed investigation of its role in glass crystallisation would be very desirable.

Other minor constituents may influence the crystallisation process because of their effects in either increasing or decreasing the glass viscosity. It should be pointed out that a given constituent can exert a greater effect than may be anticipated from its concentration in the parent glass. If the oxide in question does not participate in the structure of the crystal phase, its concentration ahead of the crystal growth front will increase, thereby giving a correspondingly greater effect upon the glass viscosity. Thus minor constituents may exert an unexpectably large effect in enhancing or diminishing crystal growth rates. Some minor constituents can effect crystal growth processes by becoming taken into solid solution in the growing crystal. The consequent changes of lattice spacings will affect the crystal growth kinetics. Clearly, further research on these questions is needed and also with regard to the extensive formation of solid solutions and of metastable crystal phases during the crystallisation of glasses.

Of particular interest is the occurrence of spherulitic growth in the crystallisation of some glasses. This form of crystallisation which characteristically occurs in viscous liquids is also important in the crystallisation of organic polymers and a valuable interaction between investigations in two fields of materials science is therefore possible.

A very important area for more extensive study, and one which is important for polycrystalline ceramics in general, concerns the relationships between the microstructure and mineralogical constitution of glass-ceramics and their physical properties. Valuable work has already been undertaken in this direction but it is clear that much of the information obtained so far is of a qualitative nature. Further research is needed, for example on the mechanical properties of glass-ceramics in which the nature of the phases and their morphology are varied in a systematic manner. Such studies would be of great value in enabling an assessment to be made of the relative importance of factors such as flaw limitation, fracture surface energy and internal stresses in determining the fracture strengths of glass-ceramics. Studies of other physical properties and the ways in which they are influenced by micro-structural variations would also be highly informative. In particular the effects on properties involving ion motion or diffusion, such as electrical conductivity or dielectric loss, merit more intensive studies. These properties appear to be highly sensitive to changes of the glass-ceramic microstructure and will therefore provide a valuable probe for investigating the microstructure. Similarly, the use of ESR (electron spin resonance) spectroscopy of glass-ceramics "doped" with suitable paramagnetic ions could provide a valuable tool for microstructural studies.

While the objective of such studies will be to establish the basic relationships between microstructure and properties, the results will undoubtedly be technologically important. If it became feasible to predict the properties of a glass-ceramic from its microstructure, the possibilities of engineering glass-ceramics for chosen applications will be facilitated.

B. Technology and Applications

The search for high strength glass-ceramics prepared from inexpensive raw materials is likely to continue. Already, glass-ceramics prepared from abundant, naturally occurring materials, such as basalt, have become firmly established in Europe for applications in which good wear-resistance is required. In a slightly different direction, the possibility exists of converting waste products into useful glass-ceramics. One such area concerns the use of slags from metallurgical processes as a basic raw material and much development of this idea has already taken place in the U.S.S.R. Another possibility is to make use of fly ash produced in large quantities by coal-burning power stations. Clearly the development of these and similar proposals is desirable not only because they will contribute to materials conservation but also because their adoption will allow disposal of large volumes of waste material for which other applications have proved difficult to find.

In addition, the development of glass-ceramics based on more conventional materials with the aim of minimising the use of expensive constituents either as major glass components or as nucleating agents would be of great benefit. Important aims will include the derivation of compositions that do not require high melting temperatures for the preparation of the parent glasses thus reducing problems of corrosion of the furnace refractory materials. Also, it would be desirable to achieve compositions having viscosity-temperature characteristics as close as possible to those of conventional glasses thus enabling a wide range of glass-shaping processes making use of existing technology to be adopted.

The achievement of low cost, high strength glass-ceramics would be of great interest to the manufacturers of containers (bottles, jars, etc.). If, by utilising the higher static and impact strengths of glass-ceramics, it proved possible to reduce the thickness and hence the weight of containers, very large savings of energy could result. The use of high strength surface-crystallised glasses in this application is a distinct possibility and in this case it is feasible that materials approaching conventional container glasses in chemical composition and viscous flow characteristics can be attained. If containers of high strength glass-based materials become available, a considerable impact on packaging technology would be expected since they could partly replace both metal cans and plastics containers. There would be distinct advantages especially in replacing plastics, because the basis of these is generally oil and conservation of this resource is desirable.

Further development of glass-ceramics for technical applications is likely to occur. For example, it has been shown that the orientation of glass-ceramic microstructures is possible and that this leads to a marked anisotropy of electrical (ionic) conductivity. This is a desirable characteristic for fast ion conductors which are needed as electrolytes for new forms of storage battery such as the sodium-sulphur battery. Thus a field exists here for the development of suitable glass-ceramics. Another requirement for such batteries is for high temperature sealants that are resistant to attack by molten alkali-metals and there is potential for development of glass-ceramics having these characteristics.

Glass-ceramics have already become established as protective coatings for metals such as mild steel but there are requirements for refractory coatings for high temperature alloys such as those used for gas-turbine rotor blades. The object here is to provide coatings that are relatively inert to inorganic contaminants present in the fuel; these can cause grain boundary attack on the alloys resulting in marked reduction of service life. Impervious coatings of high temperature glass-ceramics could provide an answer to this problem. It is perhaps questionable whether the turbine blades could be made entirely of glass-ceramic. Although the materials are strong, they are still brittle in the sense that catastrophic failure can occur once the tensile strength is exceeded. It will be recalled, however, that reinforcement of glass-ceramic with suitable fibres particularly of ductile metals results in a marked enhancement of fracture toughness and the field of glass-ceramic based composites could possibly yield a new class of engineering materials suitable for use in stringent conditions such as those in a gas-turbine engine. Even if the problem of providing suitable glass-ceramic rotor blades proves intractable, there seems to be little doubt that refractory glass-ceramics could be developed for use in other parts of the engine. Replacement of metal components by ceramic-type materials in addition to creating the possibility of higher engine temperatures and therefore greater efficiency has the added advantage of reducing the overall weight with obvious benefits in aircraft engines.

A valuable feature of glass-ceramics is that it is possible to develop microstructures containing controlled amounts of different crystal phases and of residual glass phase. Thus it is possible to produce materials having zero coefficients of thermal expansion and such materials are likely to find increasing use in sophisticated instruments. In addition, glass-ceramics for which other physical properties are practically invariant over wide temperature ranges are possible. For example, materials for which the elastic modulus is practically independent of temperature over a useful range are known and these have potential applications in new types of instruments. Another example relates to the possibility of developing glass-ceramics having a zero temperature coefficient of permittivity. Here again, the invariance of this property with temperature could be of great value in some electronics devices.

Future developments in computers may well make use of the valuable properties of glass-ceramics. Considerable effort is being expended in the reduction of the overall dimensions of computer circuitry with the aim of achieving higher computing speeds and therefore greater capacity. Thick film technology for the production of complex electrical circuits needed for interconnecting the large numbers of devices incorporated in silicon chips, requires substrate materials that are compatible with the conductor metals in several ways. As discussed earlier, glass-ceramics have excellent characteristics that enable good adhesion to metals to be achieved and this is extremely important. Also, the electrical properties of the substrate material can significantly affect the performance of the circuits and the ability to achieve controlled insulating and dielectric properties is vitally important. With thin film technology based on the deposition of conductors by evaporation techniques even higher densities of circuit elements are possible. For this technique, a substrate having a surface that is free from pits or asperities that could affect the integrity of the conductor lines is essential. The very fine grain size of glass-ceramics is likely to be a valuable asset in this connection.

In this brief summary of possible future developments in the technology of glass-ceramics, emphasis has largely been placed on the derivation and application of the materials for specialised fields. It is felt that this emphasis is justified because perhaps the most outstanding feature of glass-ceramics is their great versatility. It is probably true to say that as a group of non-metallic materials they possess greater flexibility with regard to physical property combinations than can be achieved with any group of competing materials. Thus they are capable of providing solutions to a large range of demands created by advances in many fields of engineering and it is possible that it is in these fields that glass-ceramics may ultimately have the greatest impact.

BIBLIOGRAPHY

Anon (1962). "East-West Commerce" **IX,** 9

Atkinson, D. I. H. and McMillan, P. W. (1974). *J. Mater. Sci.* **9,** 692

Atkinson, D. I. H. and McMillan, P. W. (1976). *J. Mater. Sci.* **11,** 994

Atkinson, D. I. H. and McMillan, P. W. (1977). *J. Mater. Sci.* **12,** 443

Aveston, J. (1972). *In* "The Properties of Fibre Composites", p. 63. I.P.C. Science and Technology Press, London

Baak, T. (1967). *U.S. Patent No. 3,295,944*

Badger, A. E. and Hummel, F. A. (1945). *Phys. Rev.* **68,** 231

Barry, T. I., Clinton, D., Lay, L. A., Mercer, R. A. and Miller, R. P. (1969). *J. Mater. Sci.* **4,** 596

Barry, T. I., Clinton, D., Lay, L. A., Mercer, R. A. and Miller, R. P. (1970). *J. Mater. Sci.* **5,** 117

Beall, G. H. (1971). *In* "Advances in Nucleation and Crystallisation in Glasses", p. 251. American Ceramic Society

Beall, G. H. (1972). *U.S. Patent No. 3,681,102*

Beall, G. H., Karstetter, B. R. and Rittler, H. L. (1967). *J. Am. Ceram. Soc.* **50,** 181

Beall, G. H. and Rittler, H. L. (1976). *Bull Am. Ceram. Soc.* **55,** 579

Becker, R. (1938). *Ann. Phys.***32,** 128

Becker, R. and Doering, W. (1935). *Ann. Phys.* **24,** 719

Bold, S. E. and Groves, G. W. (1978). *J. Mater. Sci.* **13,** 611

Borelli, N. F. (1967). *J. Appl. Phys.* **38,** 4243

Borelli, N. F. and Layton, M. M. (1969). *I.E.E.E. Trans. on Electron Devices,* **ED-16,** 511

Borom, M. P. (1977). *J. Am. Ceram. Soc.* **60,** 17

Borom, M. P., Turkalo, A. M. and Doremus, R. H. (1975). *J. Am. Ceram. Soc.***58,** 385

Briggs, J. and Carruthers, T. G. (1976). *Physics Chem. Glasses* **17,** 30

Brice, B. A., Halwer, M. and Speiser, R. (1950). *J. Opt. Soc. Amer.* **40,** 768

Buchi, G. J. P. and Stewart, I. M. (1962a). Private communication

Buchi, G. J. P. and Stewart, I. M. (1962b). *J. Sci. Instrum.* **39,** 487

Buerger, M. J. (1954). *Am. Mineralogist* **39,** 600

Buckle, E. R. (1960). *Nature, Lond.* **186,** 875

Burnett, D. G. and Douglas, R. W. (1971). *Physics Chem. Glasses* **12,** 117

Byer, M. (1962). *U.S. Patent No. 3,040,213*

Cahn, J. W. (1961). *J. Chem. Phys.* **42,** 93

Cahn, J. W. and Charles, R. J. (1965). *Physics Chem. Glasses* **6,** 181

Charles, R. J. and Hillig, W. B. (1965). *In* "High Strength Materials", p. 682, (Ed. V. F. Jackey). John Wiley & Sons, New York

Chyung, K., Beall, G. H. and Grossman, D. C. (1974). *Proc. Xth Int. Cong. Glass* **14,** 33

Claypoole, S. A. (1959). *Brit. Patent No. 822,272*

Cox, S. M. and Kirby, P. L. (1947). *Nature, Lond.* **159,** 162

Dalton, R. H. (1947). *U.S. Patent No. 2,422,472*

Dehoff, R. T. and Rhines, F. N. (1968). "Quantitative Microscopy", McGraw Hill, New York

Dietzel, A. (1949). *Z. Elektrochem.* **48,** 9

Doherty, P. E., Lee, D. W. and Davis, R. S. (1967). *J. Am. Ceram. Soc.* **50,** 77
Donald, I. W. and McMillan, P. W. (1976). *J. Mater. Sci.* **11,** 949
Doremus, R. H. (1973). *In* "Glass Science", p. 90. John Wiley & Sons, New York
Duke, D. A., MacDowell, J. F. and Karstetter, B. R. (1967). *J. Am. Ceram. Soc.* **50,** 67
Duke, D. A., Megles, J. E., MacDowell, J. F. and Bopps, H. F. (1968). *J. Am. Ceram. Soc.* **51,** 98
Eppler, R. A. (1963). *J. Am. Ceram. Soc.* **46,** 97
Ernsberger, F. M. (1962). *In* "Advances in Glass Technology", p. 511. Plenum Press, New York
Freeman, L. A., Howie, A., Mistry, A. B. and Gaskell, P. H. (1977). *In* "The Structure of Non-Crystalline Solids", p. 245. Taylor and Francis, London
Freiman, S. W. and Hench, L. L. (1972). *J. Am. Ceram. Soc.* **55,** 86
Freiman, S. W., Onoda, G. Y. and Pincus, A. G. (1974). *J. Am. Ceram. Soc.* **57,** 8
Frenkel, J. (1946). "Kinetic Theory of Liquids". Clarendon Press, Oxford
Galina, K. (1962). Trud 1962, Oct. 13, p. 4
Garfinkel, H. M., Rothermel, D. L. and Stookey, S. (1962). *In* "Advances in Glass Technology", p. 404. Plenum Press, New York
Gibbs, J. W. (1928). Collected Works, Vol. I. Longmans, Green and Co., New York
Goldschmidt, V. M. (1926). *Geochemische Verteilungsgesetze der Elemente,* viii, Vid. Akad.
Goss, C. L. (1962). *Prod. Engng.* **33,** 52
Gregory, A. G. and Veasey, T. J. (1971). *J. Mater. Sci.* **6,** 1312
Gregory, A. G. and Veasey, T. J. (1972). Ibid, **7,** 1327
Gregory, A. G. and Veasey, T. J. (1973). Ibid, **8,** 324
Griffith, A. A. (1920). *Phil. Trans.* **A221,** 163
Gutzow, I. and Toschev, S. (1971). *In* "Advances in Nucleation and Crystallisation in Glasses", p. 10. American Ceramic Society
Harper, H., James, P. F. and McMillan, P. W. (1970). *Discuss. Faraday Soc.***50,** 206
Harper, H. and McMillan, P. W. (1972). *Physics Chem. Glasses* **13,** 97
Has, P. D. and Stelian, L. N. (1960). *Brit. Patent No. 848,447*
Hasselman, D. P. H. and Fulrath, R. M. (1966). *J. Am. Ceram. Soc.* **49,** 68
Haward, R. N. (1944). *J. Soc. Glass Technol.* **28,** 5
Hayward, P. J. (1978). *Glass Technol.* **19,** 27
Herczog, A. (1964). *J. Am. Ceram. Soc.* **47,** 107
Heuse, E. M. and Partridge, G. (1974). *J. Mater. Sci.* **9,** 1255
Hibberd, B. and McMillan, P. W. (1975). Unpublished data, University of Warwick
Hillig, W. B. (1962). *In* "Symposium on Nucleation and Crystallization of Glass", p. 77. American Ceramic Society
Hillig, W. B. and Turnbull, D. (1956). *J. Chem. Phys.* **24,** 4
Hing, P. and McMillan, P. W. (1973a). *J. Mater. Sci.* **8,** 340
Hing, P. and McMillan, P. W. (1973b). *J. Mater. Sci.* **8,** 1041
Hollenbach, R. Z. (1963). *U.S. Patent No. 3,112,184*
Holloman, J. H. and Turnbull, D. (1951) *In* "The Solidification of Metals and Alloys", p. 1. American Institute of Mining and Metallurgical Engineers, New York
Holloman, J. H. and Turnbull, D. (1953). *Prog. Metal Phys.* **4,** 333
Hummel, F. A. (1951). *J. Am. Ceram. Soc.* **34,** 235
James, P. F. (1975). *J. Mater. Sci.* **10,** 1802
James, P. F. and McMillan, P. W. (1968). *Phil. Mag.* **18,** 863
James, P. F. and McMillan, P. W. (1970). *Physics Chem. Glasses* **11,** 59
James, P. F. and McMillan, P. W. (1971). *J. Mater. Sci.* **6,** 1401

Johnson, R. T., Morosin, B., Knotek, M. L. and Biefeld, R. M. (1975). *Phys. Lett.* **54A,** 403
Karkhanavala, M. D. and Hummel, F. A. (1953). *J. Am. Ceram. Soc.* **36,** 393
Karstetter, B. R. and Voss, R. O. (1967). *J. Am. Ceram. Soc.* **50,** 133
Kay, J. F. and Doremus, R. H. (1974). *J. Am. Ceram. Soc.* **57,** 480
Keith, H. D. and Padden, F. J. (1963). *J. Appl. Phys.* **34,** 2409
Kingery, W. D. (1960). *In* "Introduction to Ceramics", p. 478. John Wiley & Sons, New York
Kiztler, S. S. (1961). *Bull. Am. Ceram. Soc.* **40,** 231
Kokubo, T. and Tashiro, M. (1973). *J. Non-Cryst. Solids* **13,** 328
Kokubo, T. and Tashiro, M. (1976). *Bull. Inst. Chem. Res., Kyoto-Univ.* **54,** 301
Layton, M. M. and Herczog, A. (1967). *J. Am. Ceram. Soc.* **50,** 369
Layton, M. M. and Herczog, A. (1969). *Glass Technol.* **10,** 50
Levin, E. M. (1967). *J. Am. Ceram. Soc.* **50,** 29
Levin, E. M. and Block, S. (1957). *J. Am. Ceram. Soc.* **40,** 95
Levin, E. M., McMurdie, H. F. and Hall, F. P. (1956). "Phase Diagrams for Ceramists". American Ceramic Society
Levin, E. M. and McMurdie, H. F. (1959). "Phase Diagrams for Ceramists, Part II". American Ceramic Society
Levin, E. M., Robbins, C. R. and McMurdie, H. F. (1964). "Phase Diagrams for Ceramists". American Ceramic Society
Littleton, J. T. (1931). *J. Soc. Glass Technol.* **15,** 262
MacCulloch, W. T. (1968). *Brit. Dent. J.* **124,** 361
MacDowell, J. F. and Beall, G. H. (1969). *J. Am. Ceram. Soc.* **52,** 17
McCollister, H. L. and Conrad, M. A. (1966). "Fracture of Glass-ceramics", presented at Annual Meeting of American Ceramic Society
McMillan, P. W. (1970). *J. Science and Technol.* **37,** 25
McMillan, P. W. (1971). *In* "Advances in Nucleation and Crystallisation in Glasses", p. 224. American Ceramic Society
McMillan, P. W. (1974). *Proc. Xth Int. Cong. on Glass* **14,** 1
McMillan, P. W. and Hodgson, B. P. (1961). *French Patent No. 1,317,042*
McMillan, P. W. and Hodgson, B. P. (1963a). *Engng.* **196,** 366
McMillan, P. W. and Hodgson, B. P. (1963b). *Brit. Patent No. 944,571*
McMillan, P. W. and Hodgson, B. P. (1966). *Brit. Patent No. 1,023,480*
McMillan, P. W. and Hodgson, B. P. (1967). *Brit. Patent No. 1,063,291*
McMillan, P. W. and Lawton, D. C. (1967). *Brit. Patent No. 1,067,392*
McMillan, P. W. and Partridge, G. (1959a). *Brit. Patent Appl. No. 8738/59*
McMillan, P. W. and Partridge, G. (1959b). *Brit. Patent Appl. No. 8737/59*
McMillan, P. W. and Partridge, G. (1961). *French Patent No. 1,328,620*
McMillan, P. W. and Partridge, G. (1963a). *Brit. Patent No. 924,996*
McMillan, P. W. and Partridge, G. (1963b). *Brit. Patent No. 943,599*
McMillan, P. W. and Partridge, G. (1966a). *Brit. Patent No. 1,028,871*
McMillan, P. W. and Partridge, G. (1966b). *Brit. Patent No. 1,020,573*
McMillan, P. W. and Partridge, G. (1969). *Brit. Patent No. 1,151,860*
McMillan, P. W. and Partridge, G. (1970). *Brit. Patent No. 1,205,652*
McMillan, P. W. and Partridge, G. (1972). *J. Mater. Sci.* **7,** 847
McMillan, P. W. and Partridge, G. (1973). *Brit. Patent No. 1,306,727*
McMillan, P. W. and Phillips, S. V. (1962). Unpublished data, Nelson Research Laboratories
McMillan, P. W. and Tesh, J. R. (1975). *J. Mater. Sci.* **10,** 621

McMillan, P. W., Hodgson, B. P. and Partridge, G. (1966a). *Glass Technol.* **7,** 46
McMillan, P. W., Phillips, S. V. and Partridge, G. (1966b). *J. Mater. Sci.* **1,** 269
McMillan, P. W., Hodgson, B. P. and Partridge, G. (1966c). *Glass Technol.* **7,** 121
McMillan, P. W., Partridge, G. and Darrant, J. G. (1969a). *Physics Chem. Glasses* **10,** 153
McMillan, P. W., Hodgson, B. P. and Booth, R. E. (1969b). *J. Mater. Sci.* **4,** 1029
McMillan, P. W., Hodgson, B. P., Lole, J. D. and Davies, S. E. (1970). *Brit. Patent No. 1,232,621*
Maries, A. and Rogers, P. S. (1975). *Nature, Lond.* **256,** 401
Matusita, K., Sakka, S., Muki, T. and Tashiro, M. (1975). *J. Mater. Sci.* **10,** 94
Maurer, R. D. (1962). *J. Appl. Phys.* **33,** 2132
Miles, J. S. and McMillan, P. W. (1971). Unpublished data, University of Warwick
Miller, C. F. and Shepard, R. W. (1961). "Advances in Electron Tubes", (Ed. D. Slater). Pergamon Press, New York
Monneraye, M., Serindat, J. and Jouwersma, C. (1968). *Glass. Technol.* **9,** 70
Moore, H. and McMillan, P. W. (1956). *J. Soc. Glass Technol.* **40,** 66
Morell, R. (1970). Ph.D. Thesis, University of Bristol
Morell, R. and Ashbee, K. H. G. (1973). *J. Mater. Sci.* **8,** 1253
Muchow, G. M. (1971). *In* "Advances in Nucleation and Crystallisation in Glasses", p. 272, American Ceramic Society
Müller, G. (1974). *J. Dent. Res.* **53,** 1342
Olcott, J. S. and Stookey, S. D. (1962). *In* "Advances in Glass Technology", p. 400. Plenum Press, New York
Orowan, E. (1949). *Rep. Progr. Phys.* **12,** 185
Ostwald, W. (1897). *Z. Phys. Chem.* **22,** 289
Partridge, G. (1961). Private communication
Partridge, G. and McMillan, P. W. (1974). *Glass Technol.* **15,** 127
Phillips, D. C. (1974). *J. Mater. Sci.* **9,** 1847
Phillips, S. V. (1962). Private communication
Phillips, S. V. and McMillan, P. W. (1965). *Glass Technol.* **7,** 121
Pincus, A. G. (1971). *In* "Advances in Nucleation and Crystallisation in Glasses", p. 210. American Ceramic Society
Plesslinger, G. A. A., Berthold, F. and Roeder, E. (1975). *U.S. Patent No. 3,926,603*
Preston, E. (1940). *J. Soc. Glass Technol.* **15,** 262
Rawson, H. (1956). *Proc. IV Int. Congr. on Glass, Paris,* 62, Imprimerie Chaix, Paris
Réaumur, M. (1739). "Memoires de l'Academie des Sciences, 1739", p. 370
Rindone, G. E. (1962). *J. Am. Ceram. Soc.* **45,** 7
Rogers, P. S. and Williamson, J. (1969). *Glass Technol.* **10,** 46
Sambell, R. A. J., Bowen, D. H. and Phillips, D. C. (1972a). *J. Mater. Sci.* **7,** 663
Sambell, R. A. J., Bowen, D. H. and Phillips, D. C. (1972b). Ibid, 676
Sandford, E. A., Hall, D. H. and Chu, G. P. K. (1968). *U.S. Patent 3,368,712*
Sawai, I. (1961). *Glass Technol.* **2,** 243
Schreyer, W. and Schairer, J. F. (1961). *Z. Zrist.* **116,** 60
Seki, S. (1974). *Proc. X Int. Congr. on Glass* **14,** 88
Shelestak, L. J., Chavez, R. A., Mackenzie, J. D. and Dunn, B.S. (1978). *J. Non-Cryst. Solids* **27,** 83
Stanworth, J. E. (1950). *In* "Physical Properties of Glass", p. 215. Clarendon Press, Oxford
Stewart, D. R. (1971). *In* "Advances in Nucleation and Crystallization in Glasses", p. 83. American Ceramic Society

Stookey, S. D. (1947). *Brit. Patent No. 635,649*
Stookey, S. D. (1950a). *U.S. Patent No. 2,515,275*
Stookey, S. D. (1950b). *U.S. Patent No. 2,515,941*
Stookey, S. D. (1954). *U.S. Patent No. 2,684,911*
Stookey, S. D. (1956). *Brit. Patent No. 752,243*
Stookey, S. D. (1959a). *Glastech. Ber. V International Glass Congress,* **32K,** Heft V
Stookey, S. D. (1959b). *Aust. Pat. No. 46,230*
Stookey, S. D. (1960). *Brit. Patent No. 829,447*
Stookey, S. D. (1962). *Brit. Patent No. 905,253*
Stookey, S. D. and Maurer, R. D. (1962). "Progress in Ceramic Science", Vol. 2, p. 77. Pergamon Press, New York
Stryjak, A. J. and McMillan, P. W. (1978). *J. Mater. Sci.* **13,** 1275
Stutzman, R. H., Salvaggi, J. R. and Kirchner, H. P. (1959). "An Investigation of the Theoretical and Practical Aspects of the Thermal Expansion of Ceramic Materials", U.S. Dept. of Commerce, Office of Technical Services
Sun, K. H. (1947). *J. Am. Ceram. Soc.* **30,** 277
Swift, H. R. (1947). *J. Am. Ceram. Soc.* **30,** 165
Tamman, G. (1925). "The States of Aggregation". D. Van Nostrand & Co., New York
Tashiro, M. (1966). *Glass Industry* 428
Thakur, R. L. (1971). *In* "Advances in Nucleation and Crystallization in Glasses", p. 166. American Ceramic Society
Thakur, R. L. and Thiagarajan, S. (1966). *Glass and Ceramic Bull.* **13,** 33. Central Glass and Ceramic Research Institute, Calcutta
Tomozowa, M. (1971). *In* "Advances in Nucleation and Crystallization in Glasses", p. 41. American Ceramic Society
Trap, H. L. and Stevens, J. M. (1959). *Glastech, Ber.* **32K**(VI), 51
Trap, H. L. and Stevens, J. M. (1960a). *Physics Chem. Glasses* **1,** 107
Trap, H. L. and Stevens, J. M. (1960b). *Physics Chem. Glasses* **1,** 181
Turnbull, D. (1952). *J. Chem. Phys.* **20,** 411
Turnbull, D. and Cech, R. E. (1950). *J. Appl. Phys.* **21,** 804
Turnbull, D. and Vonnegut, B. (1952). *Industr. Engng, Chem.* **44,** 1292
Uhlman, D. R. (1970). *Discussion remarks, Trans. Faraday Soc.* **11B**
Uhlman, D. R. (1971). *In* "Advances in Nucleation and Crystallization in Glasses", p. 91. American Ceramic Society
Uhlman, D. R. (1972). *J. Non-Cryst. Solids* **7,** 337
Underwood, E. E. (1970). *In* "Quantitative Stereology". Addison-Wesley, Reading, Massachusetts
Utsumi, Y. and Sakka, S. (1970). *J. Am. Ceram. Soc.* **53,** 286
Vigoreaux, P. and Booth, C. F. (1950). *In* "Quartz Vibrators and their Applications". H.M. Stationary Office, London
Vonnegut, B. (1947). *J. Appl. Phys.* **18,** 593
Vonnegut, B. (1948). *J. Colloid Sci.* **3,** 563
Watanabe, M., Caporali, R. V. and Mould, R. E. (1962). *In*"Symposium on Nucleation and Crystallization of Glasses", p. 23. American Ceramic Society
Weyl, W. A. (1951). "Coloured Glasses". Society of Glass Technology
Williams, J. P., Carrier, G. B., Holland, H. J. and Farncomb, F. J. (1967). *J. Mater. Sci.* **2,** 513
Williamson, J. (1970). *Mineralogy, Mag.* **37,** 759
Williamson, J., Tipple, A. J. and Rogers, P. S. (1968). *J. Iron Steel Inst.* **206,** 898
Williamson, J., Tipple, A. J. and Rogers, P. S. (1969). *J. Mater. Sci.* **4,** 1069

Zachariasen, W. H. (1932). *J. Am. Ceram. Soc.* **54,** 3841

NB Certain Patents in which the name of the inventor is not published are referred to in the text. These are:

Brit. Patent No. 863,569 published in 1961
Brit. Patent No. 863,570 published in 1961
Brit. Patent No. 869,315 published in 1961
Brit. Patent No. 903,706 published in 1962

SUBJECT INDEX